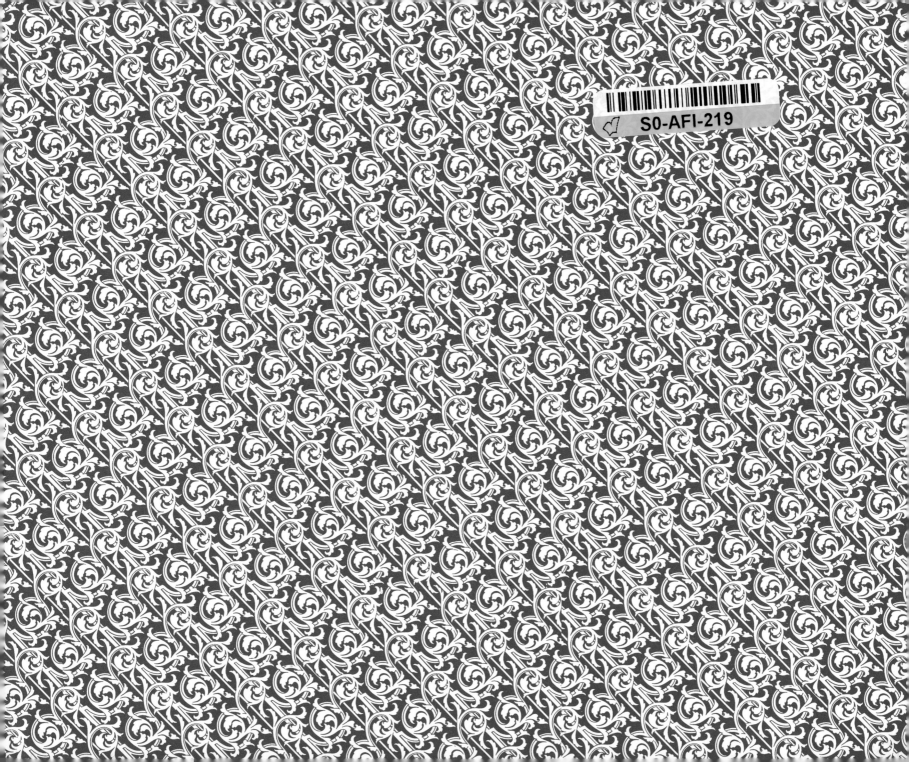

AN ASTOUNDING ATLAS OF

ALTERED ★ STATES

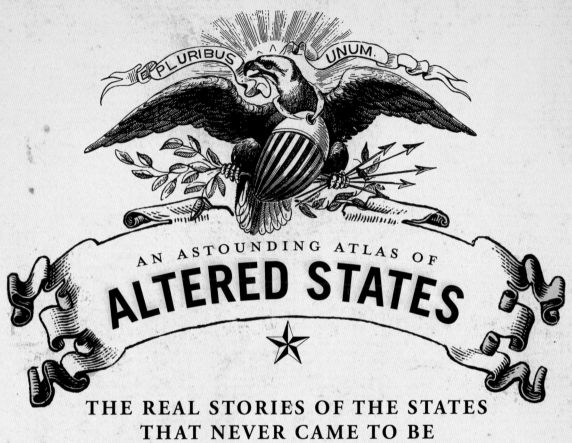

E PLURIBUS UNUM.

AN ASTOUNDING ATLAS OF

ALTERED STATES

THE REAL STORIES OF THE STATES
THAT NEVER CAME TO BE

BY MICHAEL J. TRINKLEIN

CHARTWELL
BOOKS

All images courtesy the author except the following: Chicago Historical Society, Chicago Daily News neg. coll. DN-0076962, p. 138 top; CIA World Factbook, p. 34; Library of Congress (hereafter LOC), American Memory, pp. 21, 94, 109, 118, 141; LOC, New York World Coll., p. 37, 142; U.S. Mint, p. 50; U.S. White House, p. 17.

All maps created by the author. The following resources helped in their creation: CIA World Factbook, pp. 16, 19, 55, 91, 136; (digital identification nos. are in parentheses) LOC, American Memory, p. 112 (cph 3b53089); LOC Geography and Map Div., pp. 22 (g3764b pm002791), 23 (g3764b pm002800), 24 (g3300 ct000584), 25 (g3300 ar007200), 28 (g4104c pm001550), 31 (g4050 rr001750), 33 (g4030 ct001066), 51 (g307em gct00002),52 (g3700 ct000661), 59 (g3936p rr005770), 69 (g4050 rr001760), 74 (g3701p rr000240), 75 (g4334p pm000110), 77 (g4390 ct000658), 78 (g4210 ct001350), 87 (g4030 ct001066), 97 (g3270 ct002227), 98 (g4870 ct000480), 100 (g7800 ct001786), 101 (ct001906), 105 (g3850 ct000503), 106 (g4970 ct000520), 110 (ppmsc05741), 111 (g4361h mf000060), 119 (g6710 hl000001), 120 (g3700 cw0011500), 122 (g4364l pm000250), 127 right (g3811f ar123700), 135 (g3700 ar075800), 147 (g3707o ar078901), 155 (g3870 ar139602), 156 (g3301p ct000852); LOC, Prints and Photo. Div., p. 89 (painted by Heine, J. Kummer & Döpler; LC-DIG-ppmsca-08310 DLC); Nat. Atlas of the U.S. (1970), pp. 80, 88; Oklahoma State Univ. Library, McCasland Dig. Coll., p. 115; Tom Patterson, U.S. Nat. Park Serv., pp. 12, 36, 43, 47, 48, 56, 62, 124, 131; David Rumsey Map Coll., pp 84, 103, 140; U.S. Dept. of Transp. Fed. Hwy. Admin., p. 151 (FHWA\HEPI-10, Aug. 2003).

Brimming with creative inspiration, how-to projects, and useful information to enrich your everyday life, Quarto Knows is a favorite destination for those pursuing their interests and passions. Visit our site and dig deeper with our books into your area of interest: Quarto Creates, Quarto Cooks, Quarto Homes, Quarto Lives, Quarto Drives, Quarto Explores, Quarto Gifts, or Quarto Kids.

Inspiring | Educating | Creating | Entertaining

© 2010 by Michael J. Trinklein

This edition published in 2017 by Chartwell Books, an imprint of The Quarto Group, 142 West 36th Street, 4th Floor, New York, NY 10018, USA T (212) 779-4972 F (212) 779-6058 www.QuartoKnows.com

10 9 8 7 6 5 4 3 2

ISBN-13: 978-0-7858-3452-6

Conceived, Designed and Produced by:
Quirk Books
215 Church Street
Philadelphia, Pennsylvania 19106

Previously Published as *Lost States*

Printed in China

For Lynne and Tim, who tolerate my quirks daily. And my parents, who drove Patti and me through nearly every state in the union.

INTRODUCTION	8	Long Island	64
ABOUT THE MAPS	9	Lost Dakota	67
		Lower California	68
Absaroka	10	McDonald	71
Acadia	13	Minnesota	72
Adelsverein	14	Montezuma	75
Albania	17	Muskogee	76
Alberta and British Columbia	18	Nataqua	79
(Baja) Arizona	21	Navajo	81
Boston	22	Navassa	82
Charlotina	25	New Connecticut	85
Chesapeake	26	New Sweden	86
Chicago	29	New York City	89
Chippewa	30	Newfoundland	90
Comancheria	33	Nickajack	93
Cuba	34	No Man's Land	94
Dakota	37	North Slope	97
Deseret	38	Panama	98
England, et al.	41	Philippines	101
Forgottonia	42	Popham	102
Franklin	45	Potomac	105
Greenland	46	Puerto Rico	106
Guyana	49	Rio Rico	109
Half-Breed Tracts	50	Rough and Ready	110
Hazard	53	Saipan	113
Howland	54	Sequoyah	114
Iceland	57	Shasta	117
Jacinto	58	Sicily	118
Jefferson	61	Sonora	121
Lincoln	63	South California	122

South Florida   125
South Jersey   126
South Texas   129
State "X"   130
Superior   133
Sylvania   134
Texlahoma   136
Trans-Oconee   139
Transylvania   140
Vandalia   143
Washington   144
West Florida   147
West Kansas   148
Wyoming   151
Yazoo   152
Yucatan   155

BIBLIOGRAPHY   156

★ INTRODUCTION ★

Fifty states. It's such a nice, round number. It might even seem preordained that America would gobble up the perfect amount of territory to create fifty just-right states.

Sorry. It wasn't nearly that tidy.

You know the winners, but dozens of other statehood proposals didn't quite make the cut. Some came remarkably close to joining the union. Others never had a chance. Many are still trying. It's possible that someday one of these pitches will get the green light, becoming state number fifty-one. Why stop there?

This book isn't meant to offer exhaustive detail on every unsuccessful statehood proposal; rather, the goal is to pique your curiosity, instill a sense of wonder, and enjoy a laugh or two. Because geography can be a lot of fun. Unfortunately, the study of America's geography is often framed as the memorization of state capitals. How boring! The real stories of the states are often laugh-out-loud funny, replete with absurd characters, stunning ignorance, and monumental screw-ups.

For decades I have been collecting these long-forgotten stories, and I thought it was time to share them. I've always felt a sense of wonder gazing at old maps, imagining the stories behind each squiggly line. Maps are a record of individuals trying to make a difference in how the world works. For the generation or two of Americans who have not seen any new states added, this proposition might seem odd.

After reading this book, you might see things differently.

★ ABOUT THE MAPS ★

All maps have a purpose, perhaps even an agenda. Mine are no exception.

The goal—always—was to create maps that illustrate borders and locations as clearly as possible. This means that the maps often lack extraneous detail. So don't use them to plan a hiking trip. You'll get lost.

Also, I designed each map to reflect the historic era in which the events took place. For some, I modified an existing map from the appropriate historical period; for others, I created an "old" map from scratch.

Because most failed statehood plans never had an official map, I had to conjecture a bit here and there to draw the proposed boundaries. If you have better information on a particular border, please send me a note. I'll try to include it in the second edition.

Some readers may be annoyed that I did not include inset maps that offer a "you are here" overview, but this omission was deliberate. Each map contains enough cues to address that need.

And even though certain maps may inspire you to try to establish your own state, please follow all constitutional guidelines. Don't attempt a coup. You will fail. (But if you do, please mention this book as your inspiration. It will probably help sales.)

⋆ ABSAROKA ⋆

Just what we need: Another squarish western state.

You may laugh at the notion of Absaroka becoming a state, but the same people who proposed the idea may have persuaded your family to visit this region when you were a kid.

The story goes like this: Because the area was so desolate, local businesspeople figured they needed a monumental attraction to convince people to visit.

So they carved Mount Rushmore.

Granted, I'm leaving out a lot of detail, but the point is that many of the big thinkers who pushed for the giant president heads also thought it would be nifty to have their own state.

At least that's what they proposed back in the 1930s. Absaroka would have sliced off sections of South Dakota, Montana, and Wyoming to create the forty-ninth state. The boundaries on my map are conjectural because Absaroka enthusiasts produced several different maps.

Petitions circulated, especially in South Dakota, and aggressive proponents stamped out license plates and even held a Miss Absaroka pageant in 1939. Since no follow-up contest ever occurred, I assume the winner is still wearing her tiara.

Even today, there remains a certain economic logic to Absaroka. The state would be the nation's top producer of a particularly valuable commodity: grass. There may be no better place to grow the stuff. The region's grass farmers earn a tidy income—because cows really like to eat grass. And people really like to eat cows.

Americans are also fascinated with the men who tend the cows, so that's become another economic engine for the area. For dudes who want the real cowboy experience, no other setting can match Absaroka. It's not uncommon for city slickers to spend $2,000 a day for the privilege of sleeping on the ground, drinking water from a creek, and watching horses poop (hopefully not in the same creek).

And the name? Absaroka comes from a Crow word meaning "children of the large-beaked bird." So if Sesame Street's Big Bird ever has a baby, they have to name him Absaroka, don't you think?

ABOVE --

As this fictional stamp implies, Absaroka would have had the benefit of South Dakota's biggest tourist attraction—and one of the great icons of America: Mount Rushmore. My experience in 1971 followed the pattern of most other ten-year-olds: 12 hours of "Are we there yet?" followed by 14 seconds of awe and wonder . . . followed by "What's for lunch?"

OPPOSITE --

Approximation of Absaroka's boundaries.

MONTANA
NORTH DAKOTA
ABSAROKA
SHERIDAN
SOUTH DAKOTA
Mount Rushmore
RAPID CITY
WYOMING
NEBRASKA

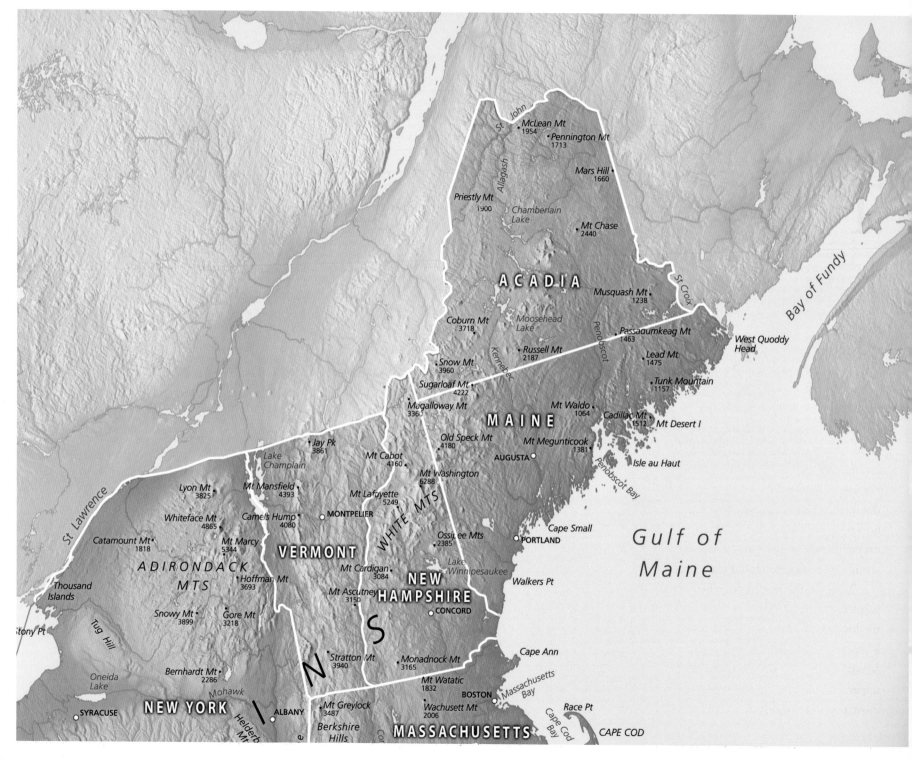

St John

McLean Mt
1954

Pennington Mt
1713

Mars Hill
1660

Allagash

Priestly Mt
1900

Chamberlain
Lake

Mt Chase
2440

ACADIA

Musquash Mt
1238

St Croix

Coburn Mt
3718

Moosehead
Lake

Penobscot

Passagumkeag Mt
1463

West Quoddy
Head

Bay of Fundy

Russell Mt
2187

Lead Mt
1475

Snow Mt
3960

Kennebec

Tunk Mountain
1157

Sugarloaf Mt
4222

Magalloway Mt
3360

MAINE

Mt Waldo
1064

Cadillac Mt
1512

Mt Desert I

Old Speck Mt
4180

Mt Megunticook
1381

Jay Pk
3861

Mt Cabot
4160

AUGUSTA

Isle au Haut

Lyon Mt
3825

Lake
Champlain

Mt Mansfield
4393

Mt Lafayette
5249

Mt Washington
6288

Penobscot Bay

Whiteface Mt
4865

Camels Hump
4080

MONTPELIER

WHITE MTS

Cape Small

*Gulf of
Maine*

Catamount Mt
1818

Mt Marcy
5344

VERMONT

Ossipee Mts
2385

PORTLAND

**ADIRONDACK
MTS**

Hoffman Mt
3693

Mt Cardigan
3084

Lake
Winnipesaukee

Walkers Pt

Thousand
Islands

St Lawrence

Mt Ascutney
3150

**NEW
HAMPSHIRE**

Snowy Mt
3899

Gore Mt
3218

CONCORD

Stony Pt

Tug
Hill

**I
N
S**

Stratton Mt
3940

Monadnock Mt
3165

Cape Ann

Bernhardt Mt
2286

Oneida
Lake

Mohawk

N

Mt Watatic
1832

BOSTON

Massachusetts
Bay

Race Pt

NEW YORK

SYRACUSE

ALBANY

Helderb
Mt

Mt Greylock
3487

Berkshire
Hills

Wachusett Mt
2006

Cod

Cape Cod
Bay

CAPE COD

MASSACHUSETTS

★ ACADIA ★

Or Just Maine. Time for a Divorce?

Today's Maine has a split personality. The south is filled with fancy folk for whom the word summer is a verb. In the north are hardscrabble Mainers living in a still-wild country of forests and mountains. Increasingly, the two groups have little in common. So in 1998, Republican representative Henry Joy sponsored a bill to study the idea of splitting the state in two.

Northerners supported the plan. They were sick of the regulations that, they believed, limit their livelihood. They want to shoot more fauna and chop more flora. Southerners, on the other hand, would prefer that everyone enjoy more civilized activities, such as growing organic blueberries or hosting Shakespeare festivals. If only the northerners could shed their genteel neighbors to the south, they could ramp up their economy by capitalizing on the resources that grow, swim, and molt throughout the region.

What would the new state be named? Many wanted to call it "Maine," which would force the lower half of the state to rename itself with a more appropriate moniker—perhaps "North Massachusetts." Others have argued that the upper half should change its name; one of the most popular suggestions was "Acadia." This name, curiously enough, applied to land that is now Maryland and Virginia in the 1500s. Over the years, the name gradually floated north until it came to rest on the region that now consists of Maine and nearby Canadian provinces.

Then there is the question of where to draw the line to form the new state. Acadia's proponents never created a definitive map showing its boundaries. I drew a fairly arbitrary east–west line. Feel free to draw your own.

Representative Joy's proposal didn't get very far, so he tried again in 2005, with the same results.

But there's always hope. Remember, the idea did work at least once before: Maine used to be just a part of Massachusetts until 1820, when it was split off to form a new state. Perhaps lightning can strike twice in the same place.

ABOVE

These loggers from Maine's north woods would probably not have been interested in getting an aromatherapy massage in Bangor. And there's the rub.

OPPOSITE

This is an educated guess of Acadia's boundary. Statehood proponents never produced an official map.

★ ADELSVEREIN ★

A New Fatherland—in Texas?

Let's move Germany to Texas." As bizarre as it sounds, that was the quite-serious plan of some rich and influential Germans in the mid-1840s.

The groundwork was laid by Gottfried Duden, whose popular German books painted an idyllic and adventurous picture of America. Even today, more Germans tour the western United States than any other European people.

By 1842 Germany's economy was failing, so twenty-one nobles devised a plan to move massive numbers of Germans to Texas. Settlers were promised comfortable travel and guaranteed jobs. The Germans were organized (of course), but a bit too optimistic. Travel costs were higher than expected, nasty weather caused problems, and disease took its toll.

Then there was the land-grant problem. In order to receive land in Texas, the Germans agreed to settle a region that was the homeland of the Comanche. I have to assume this was some sort of cruel joke, since the Comanche were known to be especially fierce. If you encroached on their territory, they'd kill you.

But Germans don't give up easily. They developed a great rapport with the Comanche and managed to strike a deal that was beneficial to both—the only time in American history that a private group forged a lasting treaty with a Plains Indian tribe. We don't know exactly why the two got along so well, but the Comanche were clearly fascinated with German leader John Meusenbach's flame-red beard—they even nicknamed him "The Red Sun." And when the toughest people on the continent give you a cool nickname, you know you've earned lasting street cred.

The Germans in Texas hoped to form their own nation, or perhaps their own state, named Adelsverein. But American state lines weren't drawn to accommodate ethnic enclaves, and the dream of a German state eventually faded.

Regardless, the lure of freedom and free land meant continued German immigration—less to Texas and more to the upper Midwest. But no one in Berlin, Wisconsin, or New Germany, Minnesota, ever proposed a new German Fatherland.

That idea never got beyond Texas.

ABOVE --------------------------------

The approximate boundaries of Adelsverein.

OPPOSITE --------------------------------

New Braunfels was the first Texas city created for German immigration. By the time this map was made, however, the dream of an independent German state had died.

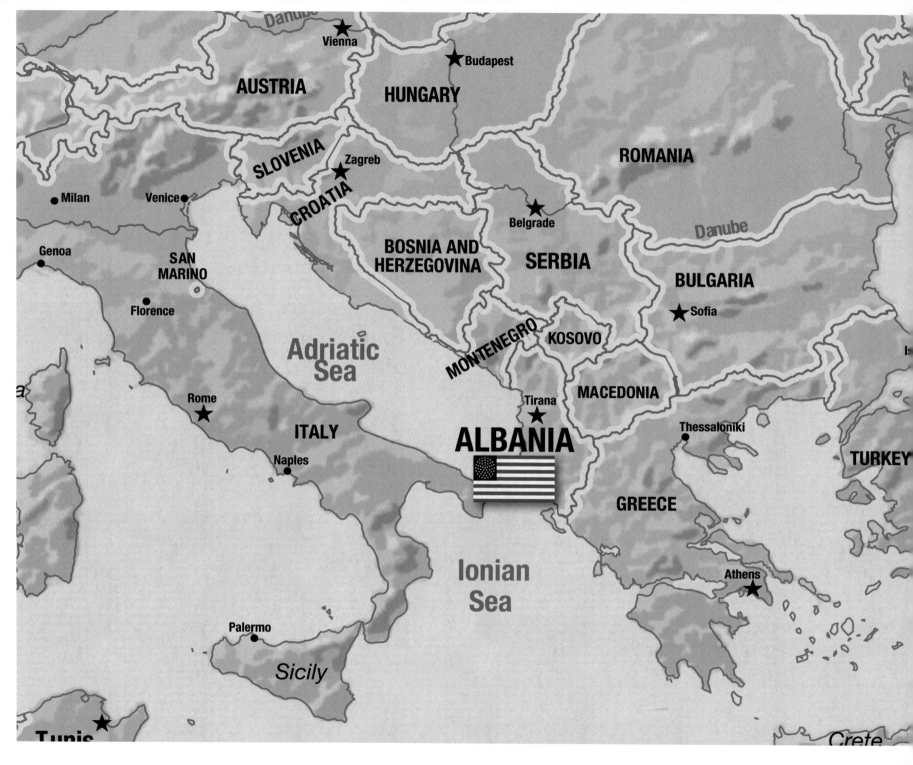

★ ALBANIA ★

They love America—like a stalker.

Thousands of Albanians would love for their country to become the fifty-first state. And it's not just some offbeat splinter group requesting the inclusion. The whole country seems rabidly pro-American.

When President George W. Bush visited in 2007, no one protested. No one. (I couldn't believe it either.) It seems that Bush enjoyed higher approval ratings in Albania than in a roomful of oil executives. In fact, his visit invoked newspaper headlines that read: "Please Occupy Us!" No kidding.

This odd love affair with America dates even earlier. During Bill Clinton's presidency, thousands of Albanians named their babies Bill and Hillary. Again, all true.

Albania was also among the first nations to join the United States in Afghanistan and Iraq, and it unflinchingly supports every American policy in the region. When the United States couldn't find any country in the world to accept deported Guantanamo detainees, Albania stepped up and took them off America's hands.

Much of this loyalty dates back to the post–World War I period, when President Woodrow Wilson made sure that the Albanian homeland wasn't chopped into sections and handed over to its neighbors. Yet that doesn't fully explain the obsession. After all, we also bailed out France in World War II, but you didn't see an uptick in French children named "Dwight" or "Franklin."

It is hard to makes sense of it all. Perhaps a little historical context is needed to understand Albania's true intentions. Remember that, during the Stalin era, Albania formed a political alliance with the USSR. When that relationship hit a rough patch, Albania jumped into bed with China, but those two lovebirds had a fight and broke up in the late 1970s. Now Albania thinks America is really cute. Right away they're bringing up statehood. That's really no different than talking about marriage on the first date.

Run, America, run!

OPPOSITE
Albania, with the 51-star flag they'd like to see.

BELOW
An unretouched photo of George W. Bush's visit to Albania. The unbridled enthusiasm is curious, but there's something even more striking about this image: Where are the women?

Picking Up the Pieces of a Broken Canada?

Canada has relationship issues. The French culture of Quebec doesn't always mesh with the English-speaking culture of the other provinces. They share a love of hockey, but that's about it.

So it should come as no surprise that Quebec has threatened to secede from Canada to form its own nation. The idea has heated up and cooled down over the years, depending on a variety of factors. In a 1995 referendum, secession lost by the slimmest margin yet: 49.5 percent to 50.5 percent. Americans may not have noticed, but their neighbor to the north came within a few thousand votes of breaking apart.

So what would happen if Quebec broke away? It's possible that the rest of the country could carry on, but a smaller, weakened version of Canada could be difficult to hold together. The strongest provinces might soon look southward for a new alliance.

Many have argued that Alberta would be most likely to petition for U.S. statehood. It has a ton of oil and a Western, "Texas-like" way of life. I imagine that Americans would love to add a new, petroleum-rich state to the union. Plus, many Albertans believe they aren't adequately compensated for all the oil flowing out

of their borders to the rest of Canada. So if the United States offered a better deal, they might just jump at statehood.

British Columbia is another possible candidate. Most of its population resides in the Vancouver area, which is just minutes from Seattle but more than a thousand miles from Calgary, its closest big-city neighbor in Canada. British Columbia also has a thriving movie industry. Partly, that's because Hollywood bigwigs think Vancouver has more of an American look than, well, America. That may be reason enough for statehood.

As for Ontario, Manitoba, and the remaining provinces, a wholesale absorption into the United States is unlikely. The rest of the world would certainly see it as a sign of U.S. imperialism and domination—a touchy subject in many parts.

OPPOSITE

One of many possible scenarios that might follow if Quebec ever becomes an independent nation.

RIGHT

There is historical precedence for British Columbia as a part of the United States. When America

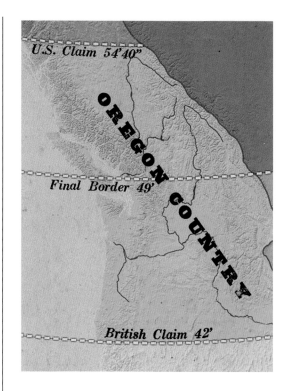

and Britain negotiated the boundary back in the early 1800s, the U.S. lobbied for a border far north of where it is today—on the southern edge of Alaska. The U.S. had more settlers in the region and threatened war if it didn't get its way. In the end, the U.S. gave in.

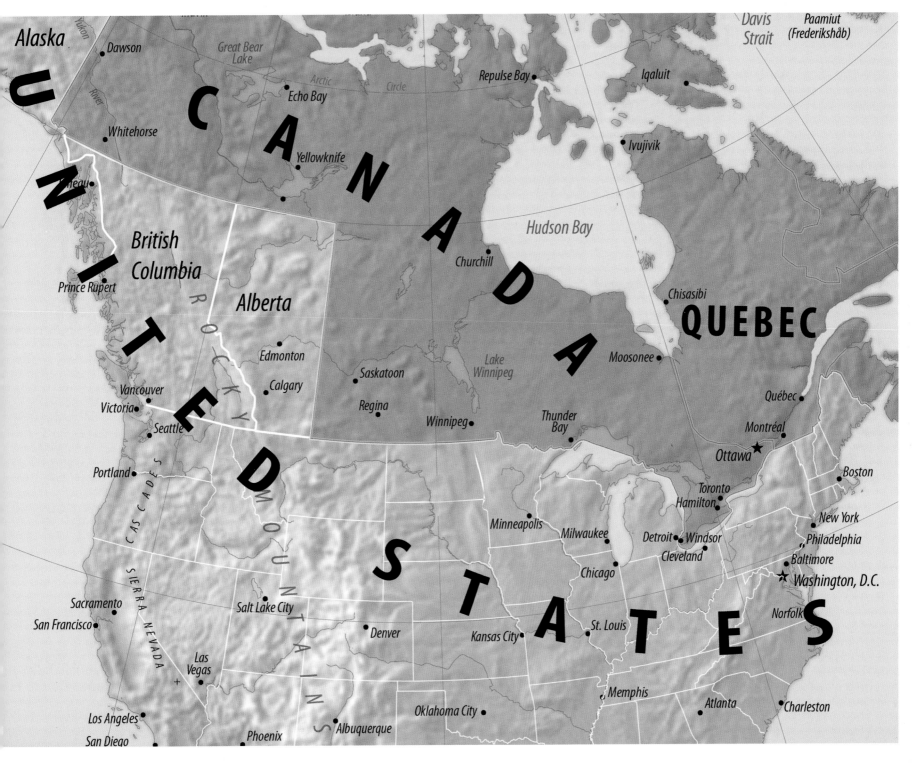

★ (BAJA) ARIZONA ★

This Is What Happens When There's No Mail.

The present border between Arizona and New Mexico was chosen out of spite. The original boundary extended east to west, not north to south as it does today. The east–west line might have become permanent had the U.S. Congress not despised the man who initially authorized it: Jefferson Davis.

The backstory goes like this: The tiny population of the American Southwest might have simply tried to ignore the Civil War, but in early 1861 the Union cut off funding for the region's only cross-country mail service. This made the locals mad. After all, interrupting mail during this era is akin to blocking cell phone service at a modern high school. Revolt was certain. In fact, Southwesterners were so upset that they decided to join the Confederacy.

Confederate president Jefferson Davis was delighted to welcome Arizona Territory into his new nation. His administration sketched out a horizontal territory that encompassed the southern half of modern New Mexico and Arizona. One of the territory's biggest benefits was that it included the best route then known for a transcontinental railroad. In fact, most folks saw no other value to the land; the notion that people might actually want to live there didn't come until much later.

As you know, the Confederacy lost the Civil War and the United States retook control of both Arizona and the Southwest. But the victors were not about to accept any boundaries drawn by Jefferson Davis or his ilk. Instead they arbitrarily chose a north–south line, thus creating the modern border between the two states.

Today, "horizontalists" still cling to the notion that southern Arizona (and maybe parts of southern New Mexico) should form a new state called Baja Arizona. The movement isn't terribly serious, but it does illuminate the distinct political contrast between the liberal Tucson region in the south and the conservative Phoenix area to the north. For example, Baja Arizona proponents acknowledge that their state would likely send two new Democrats to the U.S. Senate. Contrast that with the rest of Arizona, the state that gave rise to Republican standard-bearers Barry Goldwater and John McCain.

OPPOSITE ------------------------------

This map offers a view of the Arizona Territory as added to the Confederate States of America in the early 1860s. The territory included the Gadsden Purchase, a strip of land that provided the easiest route across the Rockies for the planned transcontinental railroad; the railway's construction was delayed by the Civil War. When work eventually started, a more central route was chosen.

BELOW ------------------------------

Today's Arizona and New Mexico border is a remnant of Northern hatred of this man: Jefferson Davis.

★ BOSTON ★

That Taxation Without Representation Thing—Again.

State Representative James H. Brennan was hopping mad. So on a hot July day in 1919, he marched over to the clerk's office in the Massachusetts House of Representatives and filed a bill to make Boston a separate state.

Brennan had his reasons. His primary beef was unjust taxation—the kind of issue Bostonians thought they'd settled 150 years before. And though there's no record of Brennan dumping any caffeinated beverages into Boston Harbor, he was just as ticked off as his tea party predecessors.

Consider this quote by Brennan: "The people of Boston must fight for the right of self-determination. The Republican Legislature [has] loaded us down with unjust taxation." He was talking specifically about $600,000 that Boston was required to pay into the state kitty for funding schools—money that would not benefit the children of greater Boston.

The whole thing blew over pretty quickly, but it points to a curious aspect of New England geography: jurisdictions divide and combine with amoebalike frequency. It started with the Pilgrims and Puritans, who split off from their homeland. The two groups were independent for a time and then melded to form the Province of Massachusetts Bay in 1691.

Maine was actually a part of Massachusetts for years, until the northerners succeeded in cleaving themselves from the state.

Then there's Roger Williams, who got along with the Pilgrims for a while, until they kicked him out in 1635. So he founded the settlement that would later become Rhode Island.

It's a good thing this trend didn't continue. If every New England squabble led to the creation of a new state, the U.S. flag would have a thousand stars.

OPPOSITE --------------------------------

The state of Boston, as perceived by James H. Brennan and his cohorts.

RIGHT ----------------------------------

Look closely at this illustration of Boston from the early twentieth century. At the bottom, you'll see the genesis of what would become the city's biggest sore spot: baseball.

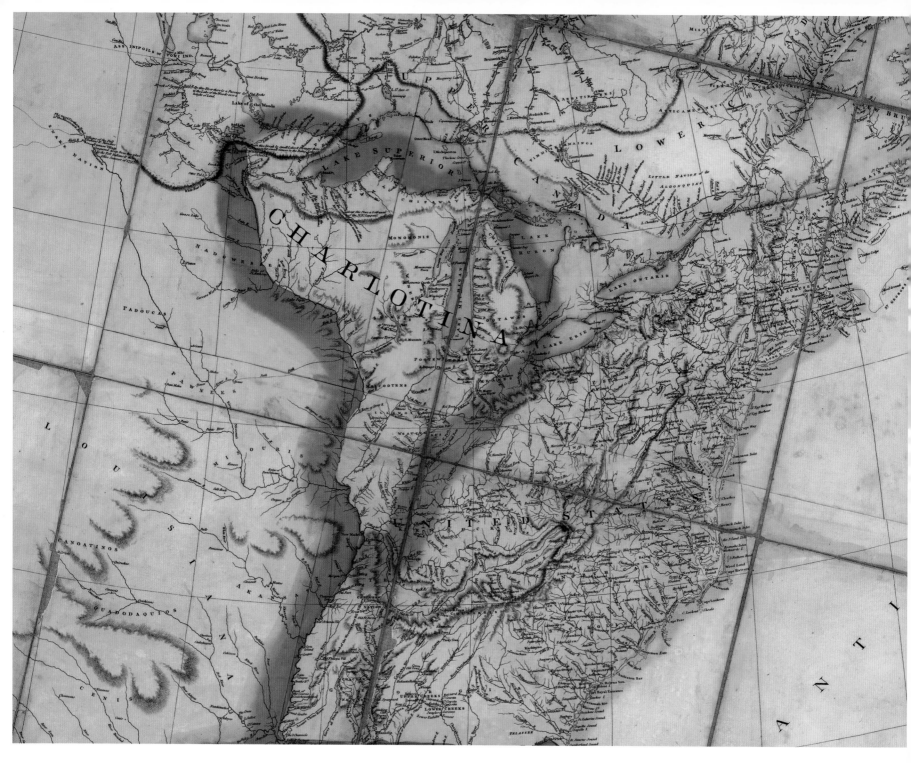

A Plan to Populate the Midwest . . . with Prisoners.

The French and British have a long history of beating up each other. So it should be no surprise that, during colonial times, their conflict spilled from the Old World into the New.

If you picture those big pull-down maps from your fourth-grade social studies class, you'll remember that in the mid-1700s, the British controlled only a portion of the eastern seaboard of the North American continent. The land between the Mississippi River and the Appalachian Mountains was claimed by their nemesis, the French.

When France and England went to war in the 1750s, it was inevitable that their American colonies would become involved. In Europe, this conflict is known as the Seven Years War; in America it became the French and Indian War. Regardless of what name you call it, the French lost, and much of their American territory was turned over to the British.

Many Brits figured that to keep their new land, they'd better get settlers there quickly, and so brochures were published extolling the virtues of a new colony called Charlotina. The proposed colony encompassed all of what is now Wisconsin, Michigan, and Illinois, plus tidbits of Indiana and Minnesota.

The pamphleteer suggested populating the new land with debtors "pining in jails throughout Britain and Ireland." As an added benefit to Britain, these new settlers could prevent a return of the French and "check Indian insurrections."

The British crown never seriously considered the Charlotina proposal. Instead, it took the opposite approach: preventing settlement in order to pacify the native peoples. The quote King George might have used (if he had borrowed from another famous George) is, "Read my lips: No new colonies." Despite the king's wishes, settlement moved forward—because people just can't resist free land.

But the borders of Charlotina were forgotten.

OPPOSITE
Charlotina overlaid on a map of the era.

ABOVE RIGHT
The colored area represents British territory in the mid-1700s, the rest is French. Even after losing their claim to middle America, French culture remained in the form of hundreds of place

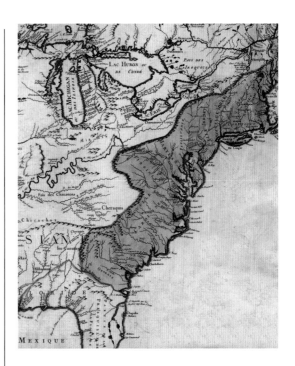

names, such as Detroit and Marquette. I think it's funny that the Germans who settled much of the region can't even pronounce the names of the cities they live in, for example, Des Plaines, Fond du Lac, and Eau Claire. Don't laugh. You try to pronounce Lake Butte des Morts.

★ CHESAPEAKE ★

A.K.A. Atlantis. A Delmarvalus Idea.

From a geographical point of view, Maryland is ridiculous. On the west side is a narrow neck that shrinks to less than two miles wide. On the eastern side, the state is split by Chesapeake Bay.

The people living on the east side of the bay complain that westsiders see them as just a "sandbox," that is, a place to play on weekends in the summer—and to ignore the rest of the year.

Then there's the issue of taxes. In the early 1970s, Eastern Shore residents became upset that their tax dollars didn't stay in their home counties—a common theme in separatist movements.

So there was talk of seceding to form a distinct state. Lots of interesting names were discussed, including Arcadia, Eastshore, Atlantis, and Chesapeake.

These early-1970s statehood proposals never involved Delaware, but some mentioned tacking on the Virginia counties that are attached to the Eastern Shore. That seemed logical—those finger-shaped counties were impossibly removed from Virginia for most of the state's history—until the completion of the twenty-three-mile Chesapeake Bay Bridge-Tunnel in 1964. (I'm not sure which is more incredible: the construction of a twenty-three-mile bridge-tunnel or the fact that they built the thing without spending a single tax dollar.)

Some Eastern Shore leaders wanted the new state to retain the name Maryland, letting their western neighbors come up with their own name, maybe West Maryland.

No discussion of this topic can avoid mentioning Delmarva, the peninsula shared by Delaware, Maryland, and Virginia. One recurring idea is to just unify the whole peninsula into one state. The proposal has been bouncing around for centuries, but it's unlikely to go anywhere because Delawareans have no interest in rethinking their state's borders. As the second-smallest state, they know they've got it good, and there's no incentive to change.

OPPOSITE

One of the 1970s plans was to split off eastern Maryland into its own state. Was it just a tongue-in-cheek proposal? It depends who you ask.

RIGHT

Because the Chesapeake Bay separates the Delmarva peninsula from civilization, residents have had to build a lot of bridges over the years. And tunnels. And bridge-tunnel combos, including one that's 23 miles long. Why not just build it all the way to Europe?

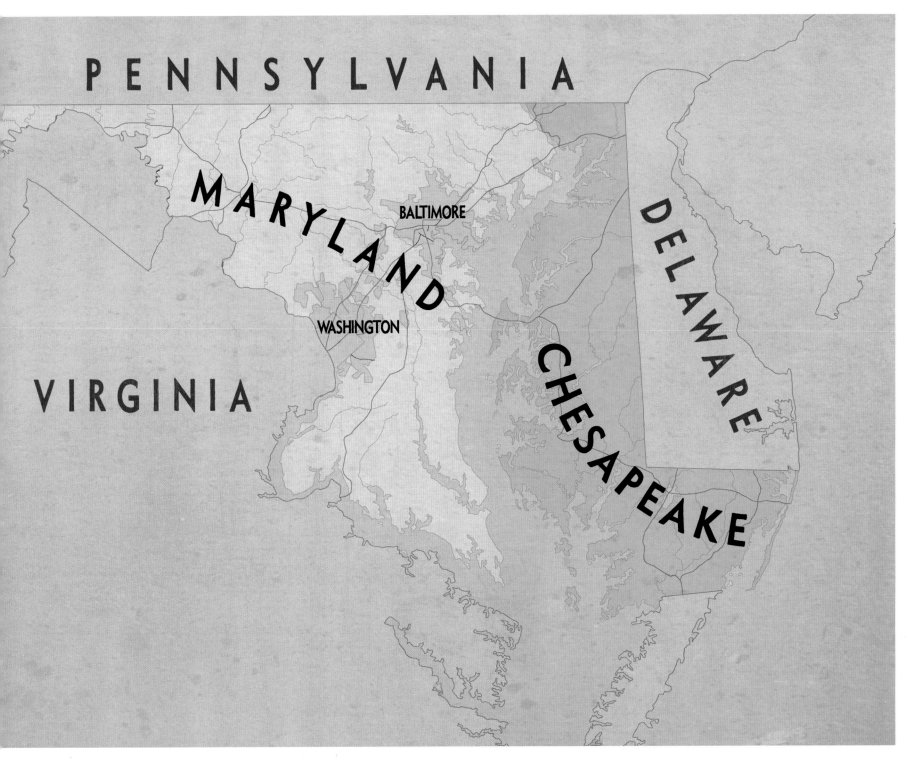

Do Farmers Control the Mafia? Kinda.

Back in the early twentieth century, Chicago was the cool place to be. Between 1900 and 1920, the city's population grew by one million. But the problem with massive growth is that it takes time for the government to catch up. By 1925, Chicago and Cook County contained almost 50 percent of the state's population but elected only 37 percent of the legislature. The new urban dwellers were none too happy that a bunch of farmers were making all the laws.

By rule, states are supposed to redistrict after each census. But the downstate farmers in Illinois refused, and somehow they managed to get away with it. Surprisingly, the Chicago Machine/Mafia cabal didn't address the matter directly by knocking some heads in Springfield, but maybe they got lost in all those corn fields. The truth: A bunch of farmers outmaneuvered the city slickers in the state legislature.

So in 1925, Chicago threatened to form its own state. Although this city-becomes-a-state thing has been popular in the Asia (think Singapore or Hong Kong), it's never been a successful venture in the United States.

Also, it's worth noting that the tension between upstate and downstate Illinois was in-tended by the folks who drew the original boundaries. If the inhabitants of the southern half had been part of some other state (such as Kentucky or Missouri), they likely would have sided with the South in the Civil War. So the boundaries were drawn (by northerners) to attach southern Illinois to a northern port.

Without this maneuver, the course of American history might have been very different. For starters, Lincoln never would have been elected to Congress if he had run for office in a pro-slavery state. And without Lincoln's ascendancy, the Civil War—if it had occurred—would certainly have had a different outcome. So thank the boundary-line creators for saving the Union. Sorta.

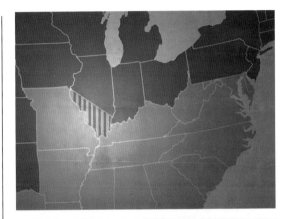

OPPOSITE

Chicago would not have covered much area, but it would have been the most densely populated state.

ABOVE RIGHT

The states that allowed slavery are in yellow. The striped area represents portions of Illinois that might have bore an allegiance to the South.

BELOW RIGHT

Abe Lincoln, in his very first photograph.

★ CHIPPEWA ★

Land of No Speed Limits.

Which way do you slice your sandwiches? Diagonally? Straight? Two pieces or four? These were the kinds of questions that Congress faced when dividing territories into states. Let's be honest, many of their decisions were arbitrary—and based on pretty flimsy reasoning.

Take the case of the Dakota Territory. Most folks thought it was too big to be one state, but no one was really sure how to slice it up. There were a lot of proposals, including the one illustrated on the opposite page. It would have created the most desolate state in the Union: Chippewa.

Chippewa is in the same vicinity as modern North Dakota, except that it's shifted west, losing the best farmland and gaining wasteland (in what is now Montana). Ever the optimists, Chippewa proponents noted the state's likelihood of getting a major east–west railroad (which the region did receive) and the tourism opportunities (during the eight weeks of the year in which there is no snow).

The problem is, no one wanted to live in Chippewa year-round. Not then. Not now. It gets nasty-cold in winter, and you can't grow anything there. True, the big sky country does evoke a certain romantic feel in the summer. That's why rich celebrities flock to the region, buy a little ranch . . . and then never come back.

Today, the biggest city in the area is Glendive, Montana (population 5,000). The town is infamous among broadcast insiders because it's the smallest TV market in the United States—and thus the worst place for an anchorman to be demoted to. A low-rated newscast in Glendive could potentially reach a viewing audience of, well, nobody.

Given the tiny population and the lack of any real economic engine, Chippewa was doomed from the start—and will likely remain that way, at least until driving really fast can somehow be turned into a profitable business. For decades, there were no speed limits on the highways because the area is so empty. How about NASCAR races on the Interstate? Now there's something you could build an economy around.

OPPOSITE --------------------------------
An approximation of Chippewa's boundaries.

BELOW -----------------------------------
Yes, it's flat and empty. That's why you can drive really, really fast.

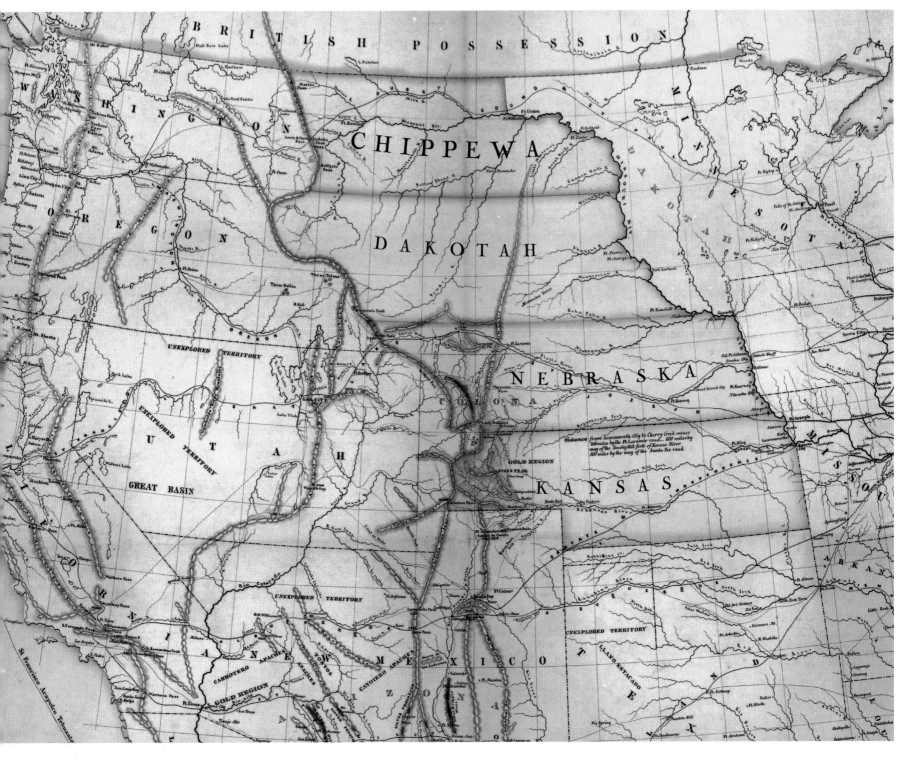

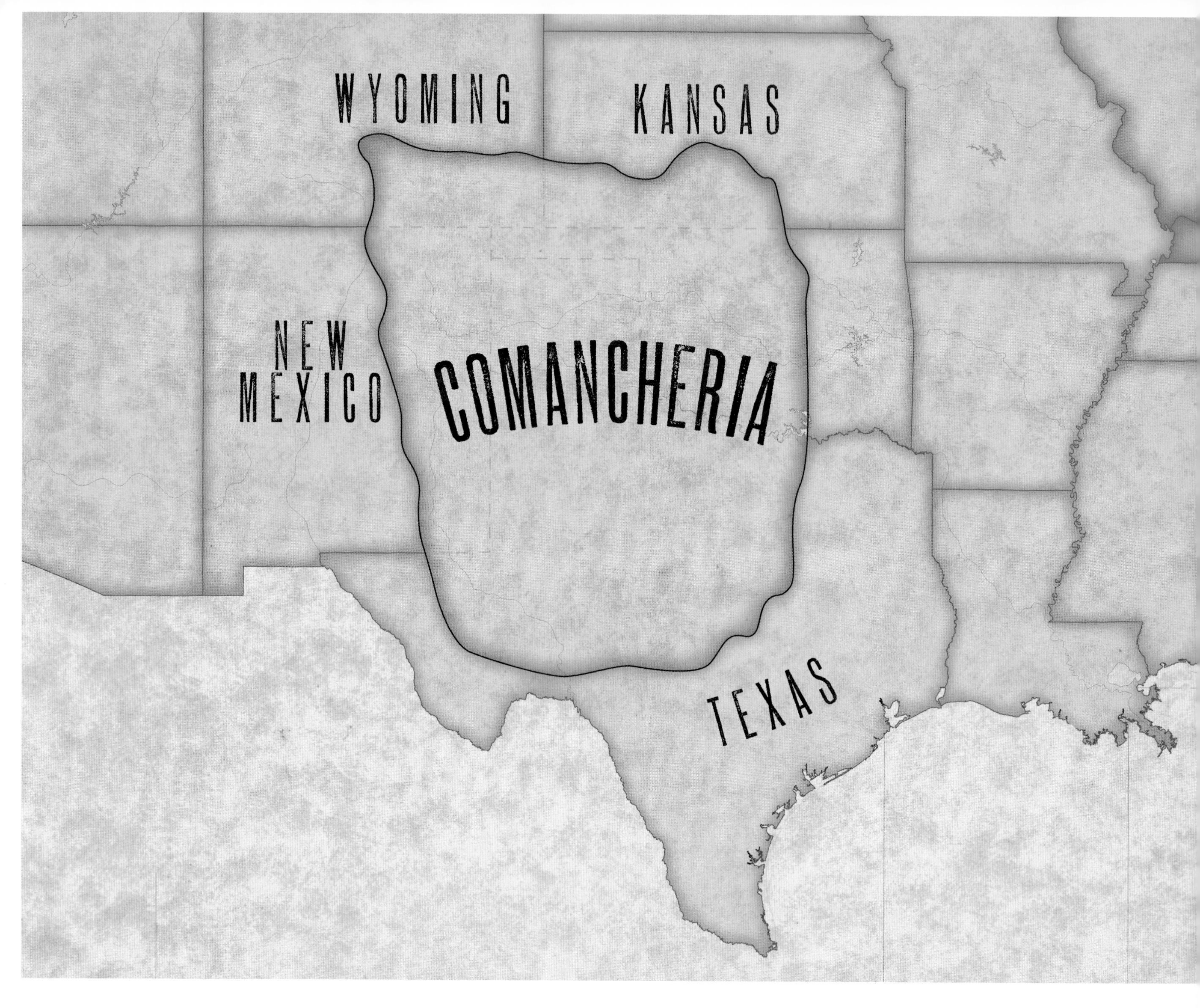
WYOMING
KANSAS
NEW MEXICO
COMANCHERIA
TEXAS

Oh, Give Them a Home . . . Never Mind, We Want It Back.

Texans are tough, but the Comanche are tougher. How else would you explain Sam Houston's willingness to cut a deal to give the Comanche people their own land?

In 1837, Texas was an independent republic. President Houston's proposal was simply an acknowledgement of what everyone already knew: The Comanche dominated much of western Texas as well as large swaths of what is now Oklahoma and New Mexico.

All kinds of dirty tricks had been used to exterminate the Comanche, including poisoning more than 350 braves in 1838. But the attempts at eradication had little effect.

Because of their incredible expertise with horses (not true of every native tribe, despite what you see in the movies), the Comanche were probably the best cavalry force in the world. It's reported that Comanche horsemen were able to ride under a horse during battle, making them nearly impossible to hit.

And let's be honest: Because they weren't afraid to make war, the Comanche were rightly feared by settlers. Of course the tribe would likely counter that they attacked only trespassers on their lands. Fair enough.

The bottom line is that many Texas leaders wanted to wall off the Comanche by setting specific boundaries. Had such a plan been ratified, it's possible that those boundaries would have later become the outline of a new state.

The Texas congress did not ratify Houston's plan. They hoped that, given enough time and enough new settlers, they could simply overpower the Comanche. In the end, that proved true.

OPPOSITE

Comancheria, as incorporated into the boundaries of a modern map.

RIGHT

The empty areas in this west Texas map were Comanche territory—whether the Texas government acknowledged it or not.

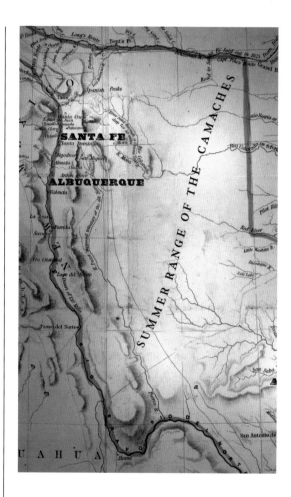

★ CUBA ★

A Serious Proposal That Might Have Saved Civilization.

It might seem ludicrous to claim that a proposal to make Cuba into a state had the potential to save the earth—but I'm writing this with a straight face.

In case you aren't old enough to remember the Cuban Missile Crisis in 1962, here's a primer: The Soviet Union decided to park some nuclear missiles in Cuba and point them at U.S. cities. This was bad, because they could then blast the United States into the Stone Age before America had a chance to return the favor.

Politicians call this type of thing "destabilizing," but for many Americans at the time, a better word was "panic." All-out nuclear war seemed increasingly possible. Thankfully, U.S. president John Kennedy deftly defused the situation, and the Soviets removed their missiles from Cuba. A nation exhaled.

But the United States wouldn't have come to the brink of annihilation if an earlier statehood proposal had taken root. A few decades earlier, many influential Americans suggested that Cuba should become a state.

The idea dates back to 1898 and the end of the Spanish-American War. That summer, Spain surrendered and agreed to relinquish the islands of Puerto Rico, Guam, the Philippines, and Cuba. The United States took possession of the first three, but decided to give Cuba its independence (although the United States did retain Guantanamo). Nationhood went well for a few years, but when the Cuban president decided to overstay his term, revolt and turmoil erupted across the island. Many in the U.S. Congress saw annexation as a solution. The New York Times polled influential politicians and found many in favor. Senator Stephen B. Elkins of West Virginia was among those who thought Cuba should be annexed, much as Texas had been. On the opposing side, Representative John Sharp Williams of Mississippi was against the idea because, in his words, "we have enough people of the Negro race."

Even at that time, Cuba's proximity to the United States was understood as vital to America's national defense. (Cuba is so close to Florida that marathon swimmers can make it from one coast to the other). Nonetheless, it appears racism was the primary reason for denying Cuba's entry into the union in 1902.

And that brings us to 1962. If Cuba had been American soil, there would have been no Cuban Missile Crisis—one fewer near-death experience for the planet.

ABOVE --

Photographed by U.S. spy planes, this image was one of many that proved the Soviets were setting up nuclear devices in Cuba.

OPPOSITE --

Government maps such as this one scared Americans to death in 1962. In minutes the Soviets' MRBMs (medium-range ballistic missiles) stationed in Cuba could reach Washington, not to mention a vast swath of the American Southeast. IRBMs (intermediate-range ballistic missiles) could reach pretty much everywhere else.

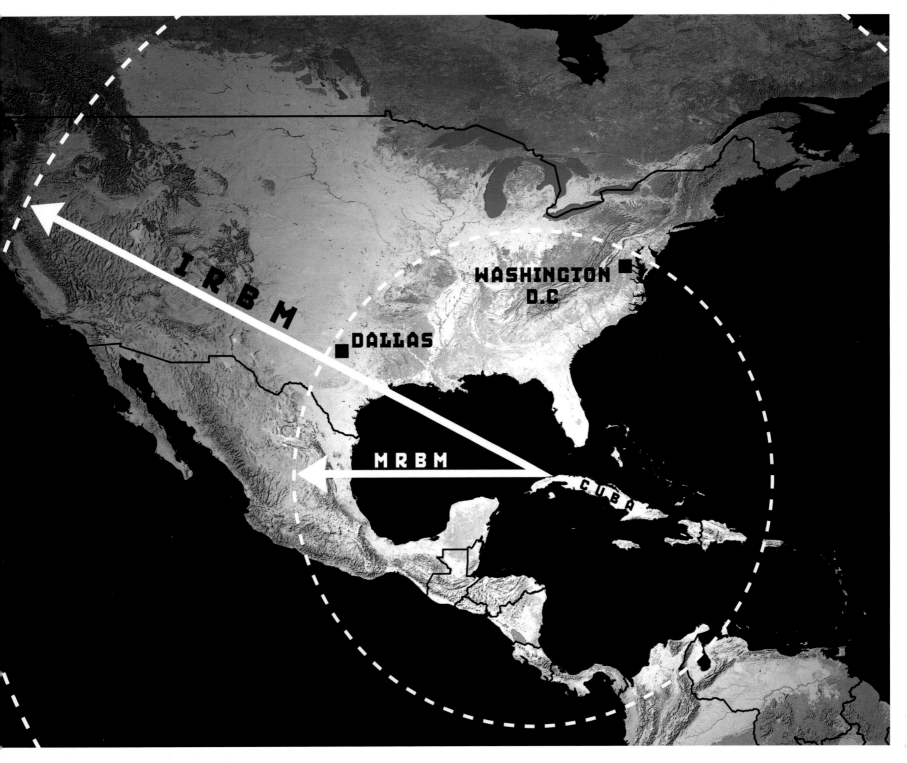

DAKOTA

DAKOTA

SOUTH
DAKOTA

WEST
DAKOTA

EAST
DAKOTA

PEMBINA

DAKOTA

★ DAKOTA ★

Or Pembina. A Genius Marketing Idea.

Which of the four proposals shown here for the Dakota Territory finally became law? The answer—which you should know—is none of the above. All four failed, but it wasn't for a lack of trying.

The very first Dakota statehood bill suggested that the region be one giant state. The idea received significant support, but then legislators apparently calculated how long it would take to ride a horse from the state capital to the outlying counties. Heck, even at today's interstate speeds, crossing one of the Dakotas seems to take an eternity.

The 1877 plan to divide the territory into eastern and western states actually makes sense, geographically. The east has good farmland and precipitation; the west doesn't. The problem, of course, is that West Dakota would have no people. States need people.

The 1882 plan would have made the southern half into a state—named Dakota—and designated the northern half as Pembina Territory. This proposal had popular support, but the nation's Democratic president wasn't about to add any new Republican-leaning states.

But in 1888, Benjamin Harrison, a Republican, won the presidency, and within months he signed a bill admitting the states of North Dakota and South Dakota into the Union.

The story doesn't end there. Leaders in North Dakota periodically trot out the notion that north is a pejorative term; they suggest changing their state's name to just plain Dakota. This idea is proposed with some regularity (namely, in 1947, 1983, 1989, and 2001), and it gets news coverage every time.

I have to admit, plain "Dakota" is a great name. Dodge even adopted it for their tough, outdoorsy truck. And anytime an automaker stamps a state or city name on a vehicle, it's because focus groups love it. Examples include the Pontiac Montana, Chevy Malibu, Toyota Tacoma, and GMC Yukon. On the flip side, I don't think you'll ever see a Subaru Saskatchewan or an Acura Arkansas.

OPPOSITE
Which of these four proposals became law?

RIGHT
North Dakota could cash in on the naming-rights craze by securing automaker Dodge as a sponsor and then renaming the state "Dodge Dakota." As preposterous as that sounds, there is a real-life precedent. In 1950, Ralph Edwards persuaded residents of Hot Springs, New Mexico (pop. 4,200), to rename their city after his "Truth or Consequences" radio show. The city remains Truth or Consequences to this day. Check your map.

★ DESERET ★

That's Deseret, Not Desert. The American Theocracy.

The Mormon church was born in New York in the 1820s. So how did it end up in Utah? Almost from the beginning, the church's unusual beliefs led to persecution, and members kept transplanting themselves to avoid harassment.

Things got especially messy when founder Joseph Smith received a revelation that he should start taking more wives. Lots more. Soon he was encouraging other Mormon men to do the same. This diminished the available supply of comely young women, which upset a lot of non-Mormon men, especially the bachelors. For this reason—among others—an angry mob killed Smith in 1844.

This left master-organizer Brigham Young in charge. He solved the angry-bachelor problem by deciding to move the Mormons west to a place that had no white settlements. Thus began the Mormon exodus to the Salt Lake basin in the mid-1840s.

It wasn't long before Young petitioned Congress to create a new state for his people. Although the Mormon leader wanted a giant property, he skillfully drew boundaries to avoid conflicts with established outposts such as the California gold fields or Oregon's newly pop-

ular Willamette Valley. And he made sure to grab a bit of coastline in southern California. At this point in history, southern California was pretty empty. Hard to imagine.

Young's super-sized state never came to be, mostly because of an anti-Mormon bias then pervasive in American culture. (This was long before the Osmond family taught us that Mormons are super-sweet folks with great teeth.) After a few decades, the Mormons abandoned polygamy (sort of), and Congress finally offered statehood.

Citizens overwhelmingly preferred calling their new state Deseret, a name found in the Book of Mormon. But the federal government chose Utah, after the Ute tribe.

The state's final size was much smaller than the Mormons had originally hoped. The entire western half of the Mormon empire was sliced off to form Nevada, establishing one of the oddest geographical juxtapositions in the United States. Could two states like Nevada and Utah be more different? Is it a joke that they are adjacent? In Utah, it's really hard to get a drink. In Nevada, it's really hard not to get a drink.

OPPOSITE

The approximate boundaries of Young's Deseret.

BELOW

Brigham Young laid out a perfect grid for Salt Lake City. If you visit, you'll immediately see how ridiculously wide the streets are. Young wanted to ensure that a team of horses could make a U-turn anywhere.

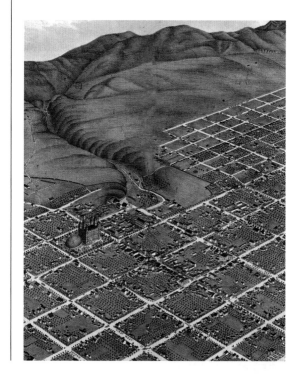

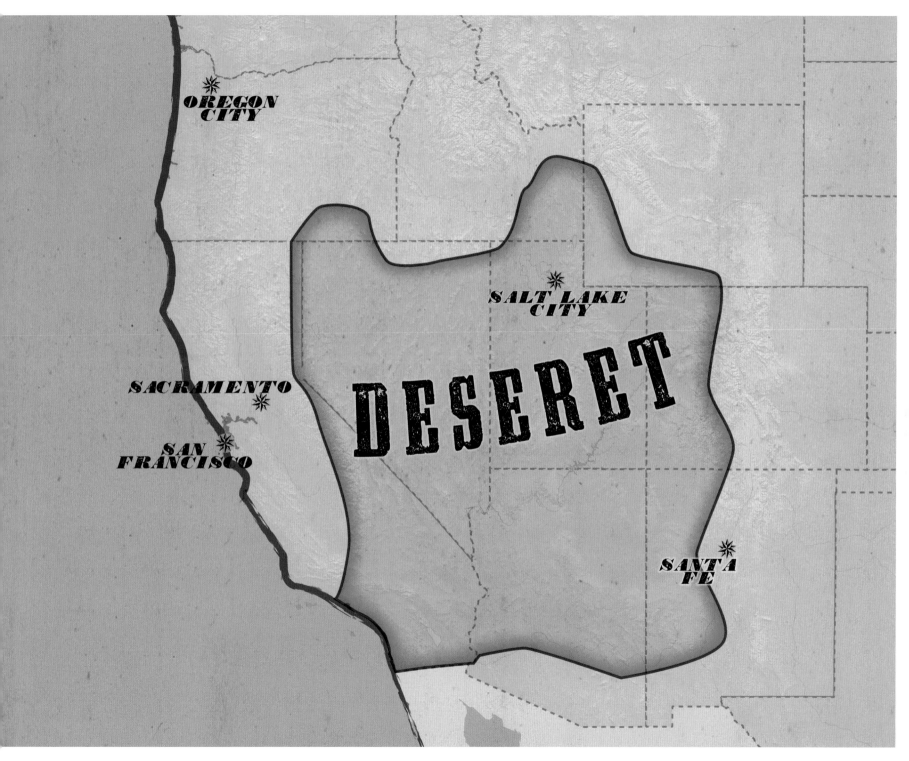

ALL THE NEWS
THAT MATTERS

The Daily Herald

LATEST
EDITION

VOL LXII   MARCH 4, 1947   DAILY 5 CENTS

ENGLAND, SCOTLAND, IRELAND, WALES JOIN UNION

As states #49, #50, #51, #52

First new states since Arizona, New Mex.

Role of King Unclear as Nations come Together

LONDON - UPI - In a surprise move, the British people today approved a referendum to join the United States. England will become the 49th state, Wales, the 50th, Scotland the 51st and Ireland the 52nd. The move was applauded by many including the U.S. president Harry Truman. Winston Churchill also was in favor of the plan.

However the status of the British monarchy is not yet clear. The king has not spoken on the matter. The previous king abdicated, and then married an American divorcee, but that scandal is said to have no bearing on the statehood matter.

Statehood was first proposed by Sen. Russell of Georgia. The idea was first received coolly. When Sen. Russell pointed out all America had done for Britain in WWII, many in England responded that the U.S. still owed money to Britain—money borrowed to pay for the Civil War.

(continued on page 16)

Frosts Predicted

lation bookings at the City Jail gets under way gas generated Shipyard Workers

IN THE /TIMES/ TODAY

MAY REFUSE POST
Beyond question Giraud was re-

★ ENGLAND, ET AL. ★

A Bloody Awful Idea. But You've Got to Love the British Comeback.

America should add England, Scotland, Ireland, and Wales as four new states—or so proposed Sen. Richard Russell Jr. of Georgia. He may have been the only person on earth who thought that was a good idea.

Normally, this kind of thing would be laughed off as the idea of a crackpot, but because the plan was proposed by a U.S. senator, the media dutifully reported on it. Press reports from 1947 note that the British received the idea "coldly."

How coldly? Think absolute zero. If you skipped high school chemistry, absolute zero is the coldest known temperature. In theory, nothing in the universe can actually reach absolute zero—but I think maybe the British did in 1947.

Of course, they didn't just launch into a tirade of expletives. The British are much too sophisticated for such a coarse response. So rather than condemn Senator Russell, plucky Brits simply pointed out that Georgia still owed money borrowed from the British during the Civil War.

It was the perfect retort. Russell assumed that America held the upper hand since the United States had just bailed out Britain in World War II. He had forgotten that England had spent millions to help the South in the war between the states—money that was never repaid.

Hmm. Maybe this whole thing was backward. Perhaps Georgia should be added to the United Kingdom.

As the rhetoric escalated, Southerners claimed they really didn't owe anything because the Civil War debts were payable in Confederate dollars. Since the Confederacy didn't exist, they were off the hook. It's as though the South had hired one of those late night TV bankruptcy lawyers.

It's a loophole Americans should keep in mind as the national debt escalates. All the United States has to do is pick a fight with Canada and then lose the war. When debtors try to collect, America responds with, "Sorry, we owed you American dollars. All we have now is this goofy-looking Canadian money."

Fortunately, after a brief flurry in the spring of 1947, the whole U.K. statehood idea died an appropriate death.

OPPOSITE

The headline Sen. Richard Russell was hoping for. Of course, it never happened—and it wasn't his only lost cause. Later, he thought he could win the presidency. That didn't work out either.

BELOW

Sen. Richard Russell (right), with Pres. Lyndon Johnson.

★ FORGOTTONIA ★

An Accurate Name for the Land Illinois Forgot

Illinois is pretty much all corn. Once you venture outside Chicago, the state becomes one giant field dedicated to the creation of tasty high-fructose corn syrup. To serve the corn farmers, small and medium-size towns have grown up all across the Land of Lincoln—cities like Peoria, Decator, and Springfield.

What's curious is that interstate highways connect nearly every one of these villages. Compare a map of Illinois to a map of, say, Iowa or Missouri, and you almost have to laugh: Illinois has so many more miles of freeway compared to its neighbors.

Here's how it happened: A big city like Chicago has a lot of people, and that means a lot of money and a lot of representation in Congress. But you can build just so many highways in the Windy City. So the "downstate" residents reap the benefits of living in a populous state. How else can you explain a freeway that has the singular purpose of connecting Galesburg and Moline?

Having said all that, one section of the state has been left out: the counties in the western bulge. Largely cut off by the Illinois River, this area didn't get any fancy freeways in the 1960s and 1970s.

In protest, a group of residents decided to form their own state, Forgottonia. They appointed a governor and tried to attract attention. But what they really wanted was an interstate. Specifically, Interstate 72, which would provide a shortcut between Chicago and Kansas City. Legislation that would have authorized the construction of I-72 was defeated in Congress in 1968 and again in 1972. Parts were eventually built decades later, but even today I-72 extends only to the Illinois–Missouri border.

And so Forgottonia still struggles. Businesses have steadily left. Amtrak's arm had to be twisted to ensure continued service. The region even had a college up and move to a different state, which is pretty amazing considering the infrastructure they left behind.

Such is the sad story of Forgottonia. It never had a real shot at statehood—and it's still pretty much forgotten. But they do have corn. Lots and lots of corn. So as long as America keeps drinking sixty-four-ounce fountain drinks, Forgottonia's people will survive. In fact, about the only thing that could hurt Forgottonia would be medical reports suggesting high-fructose corn syrup isn't healthy.

Oh.

OPPOSITE

Forgottonia superimposed on a map of the Midwest. The original statehood proposal included 14 counties; this map scales back to the 10 counties on the west side of the Illinois River.

BELOW

It's corn syrup. If you assumed the barrels were filled with toxic waste, you'd be wrong. Sort of.

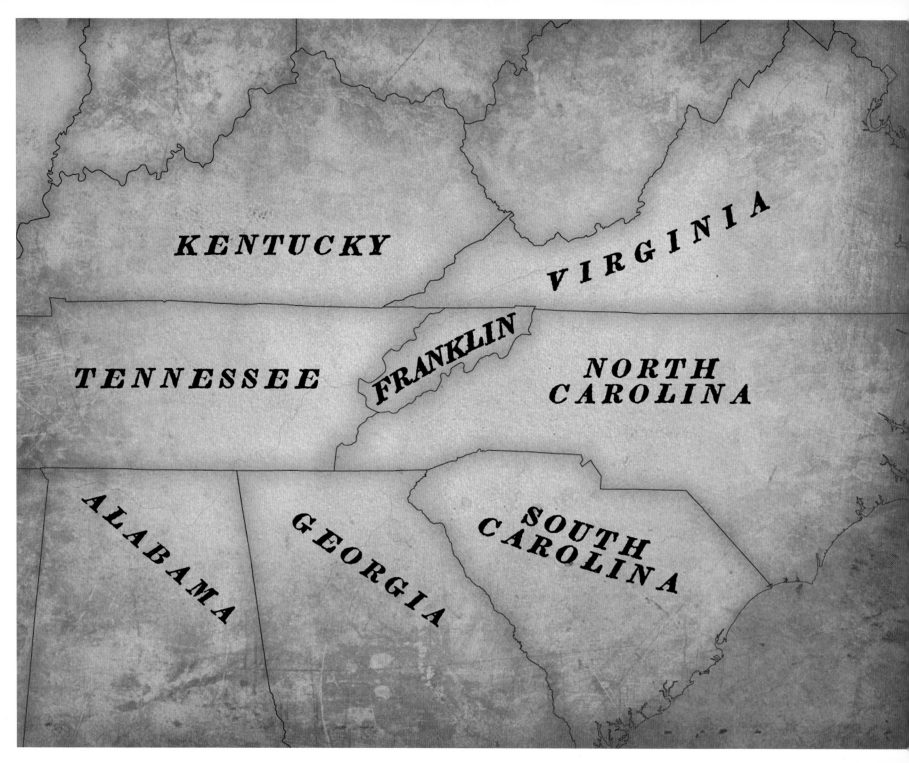

★ FRANKLIN ★

First, We Kill All the Lawyers.

What were they thinking? That's the question I keep asking myself about the frontier folk of Franklin. Back in 1785, residents of western North Carolina (now eastern Tennessee) tried to organize themselves into a new state. Things started downhill with their proposed constitution: It banned doctors and lawyers from election to the legislature. (These hardscrabble folks had a real bias against their gentrified neighbors.)

More than half the existing states voted in favor of admitting Franklin to the union, but the vote fell short of the necessary two-thirds majority.

Franklin proponents then figured they needed a public relations campaign to win the remaining votes, so they tried to enlist the nation's most respected elder statesman: Ben Franklin. After all, they had named the state after him. But Ben turned them down flat. The lesson here is that if you're going to name a state after a living historic figure, you should probably ask the person first.

Undeterred by all the rejection, the people of Franklin decided to pretend they were a state anyway. They elected a government and established courts. And in a popular move, they decided not to tax anyone for at least two years. That sounds pretty good . . . until you try to raise a militia.

And Franklin needed a militia. Badly. As a rogue state in the late 1700s, they didn't enjoy the protection of the federal government, and it wasn't long before local Native Americans figured out that Franklin's settlers were sitting ducks. Attacks soon escalated.

Desperate for help, Franklin's nascent government tried to become part of Spain. Seriously. And it's not quite as bizarre as it might seem. Spain was a major power in North America (remember Florida?) and had great wealth to boot. But appealing to Spain really ticked off the leaders of nearby North Carolina. They didn't want some foreign country setting up camp on their border. So soldiers from North Carolina marched over and arrested the governor of Franklin.

That pretty much put an end to the state of Franklin experiment. The area was folded into the new state of Tennessee in 1796.

It's worth noting that Congress had an unofficial policy against naming new states after people (dead or alive). It's a guideline they held for more than a century, with only one exception: Washington.

OPPOSITE

Proposed state of Franklin superimposed on modern borders.

BELOW

Franklin didn't like Franklin.

WOW, THIS IS A BAD IDEA!

★ GREENLAND ★

Prime Real Estate. Even George W. Bush Was Interested.

Location, location, location. Those are the three main reasons the United States has been trying to acquire Greenland since the mid-1800s. But the owner—Denmark—isn't selling.

Sure, Greenland's a fixer-upper, but America is still willing to pay full price for the huge chunk of oceanfront property. That's because Greenland sits directly between the United States and Russia—prime location for a missile defense system.

The first real estate agent who tried to broker a sale of Greenland was none other than Secretary of State William Seward. And although he couldn't quite close the deal on Greenland, he did manage to arrange the purchase of Alaska for 1.9 cents an acre. Some criticized Seward for overpaying, but in retrospect he made a monumentally good deal, buying a whole state for about the cost of a nice house in southern California today ($7.2 million to be exact—the going rate for a home suitable for a B-list actor).

Seward has been largely forgotten, though I'd argue that he's one of America's great unsung heroes. He worked tirelessly for the rights of the mentally ill, deeply opposed slavery, and fought for prison reform. On the night Abraham Lincoln was assassinated, a similar attempt was made on Seward because he was the president's secretary of state. He bore the deep scars of the attack—on his face—for the rest of his life.

Eighty years after Seward, a new secretary of state, James Byrnes, offered $100 million to the Danes for Greenland, but he didn't even elicit a counteroffer.

More recently, President George W. Bush went to Denmark to appeal to the Danish government. He wasn't looking to buy Greenland outright; instead, he was laying the groundwork for something of a co-op deal. Denmark would retain the title to the land, but America wanted the right to set up its military stuff. Nothing official was signed since these deals seem to take decades to work out.

If the United States ever does purchase Greenland, the prospect of the island becoming the fifty-first state isn't quite as dim as you might expect. It's true that Greenland is mostly ice, but that could change quickly as global warming kicks in. Just a few degrees difference could make Greenland a much more hospitable place to live. But I don't think Hawaii has to worry about losing its pineapple industry—yet.

ABOVE --------------------------------

The sign America has been hoping for.

OPPOSITE --------------------------------

About the only way to understand the value of Greenland is to see it from space. It occupies a prime location between Russia and North America.

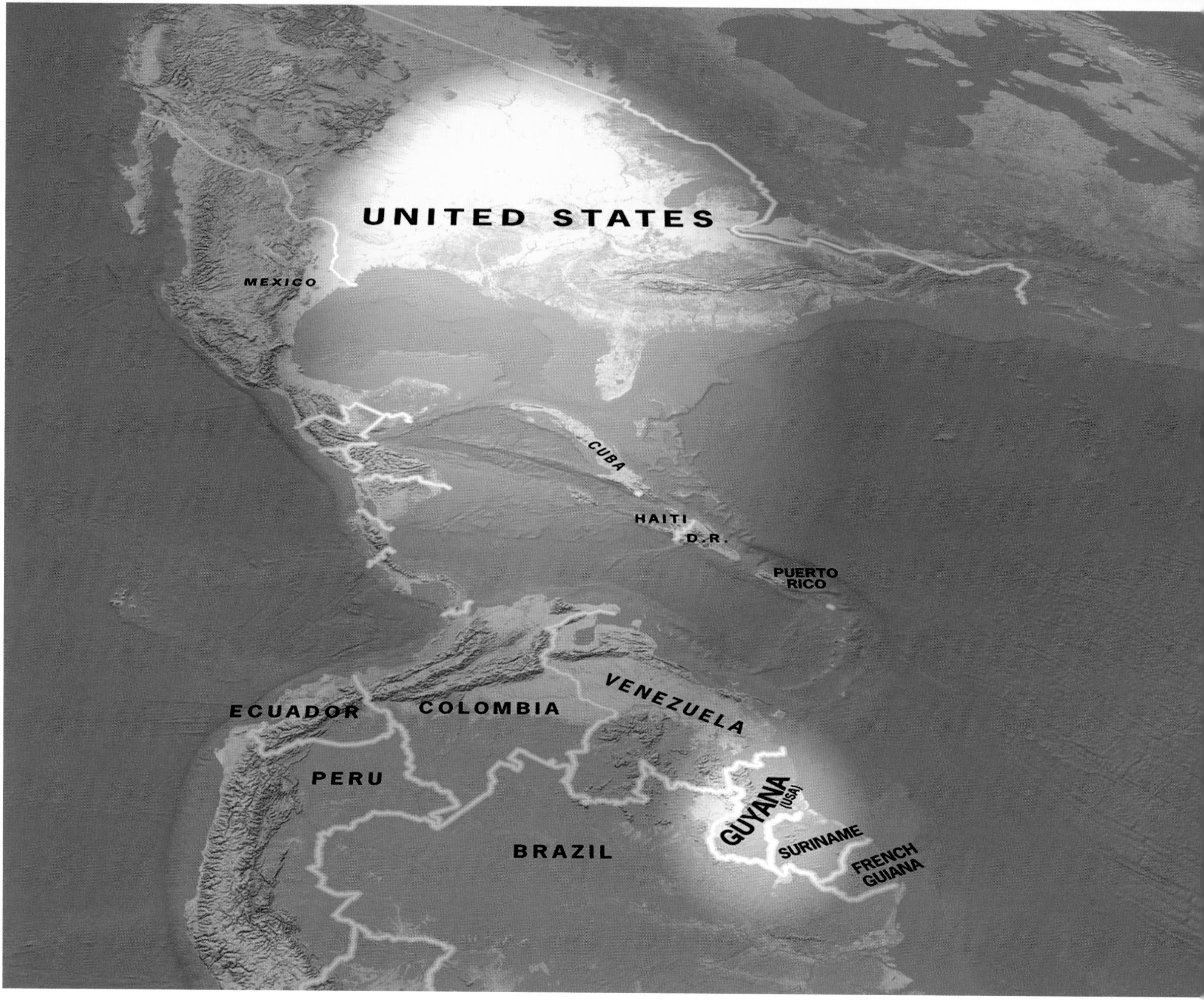
UNITED STATES
MEXICO
CUBA
HAITI
D.R.
PUERTO RICO
VENEZUELA
ECUADOR
COLOMBIA
PERU
GUYANA (USA)
SURINAME
FRENCH GUIANA
BRAZIL

★ GUYANA ★

An Idea That Makes NASA Drool.

Most Americans haven't thought much about Guyana since the very disturbed Jim Jones wiped out his cult there in 1978.

These days, a group called GuyanaUSA wants to push the country back into the headlines, but this time the message is more positive. Kind of.

Guyanans love America—maybe too much. The country is essentially emptying as all its residents move to the United States. If the trend continues, the last one to leave can turn off the lights in just a couple decades.

Even if emigration slows, Guyana is already in danger of collapse. The country has lost many of its skilled workers. If it ever disappears from the map, the land would likely be divided among neighbors Brazil, Venezuela, and Suriname.

GuyanaUSA believes that the only solution is U.S. statehood, or at least commonwealth status, similar to Puerto Rico. Before you dismiss the notion, consider a few intriguing facts that suggest Guyana might make a good fit with the United States after all.

First, because the country was formerly a British colony, everyone speaks English. In fact, Guyana has a higher percentage of English speakers than the United States (99 percent in Guyana versus 85–96 percent in the U.S., depending on whose statistics you cite). The Guyanese people also have a higher literacy rate than Americans: 96 percent, according to recent statistics. And the majority of the population is Christian, just like the United States.

It's also worth noting that Guyana's biodiversity rivals any on earth. Many have argued that the best way to preserve its rain forests is to offer the protection of U.S. law. (Admittedly, this is not a terribly compelling reason to annex another country, but America has invaded places for flimsier reasons.)

Finally, there is the avid support of rocket scientists. Apparently, Guyana is a great place to launch spaceships—much better than Florida. I don't fully understand the physics (because it's, well, rocket science), but Guyana's proximity to the equator means that launches can take advantage of the earth's rotation and get something akin to a running start.

Despite all these factors, statehood for Guyana is the longest of long shots. The U.S. government doesn't want to be viewed as a colonizing power. The poverty level and lack of infrastructure would require a tremendous capital infusion. And the current government in Guyana has no interest in giving up power.

Then again, Guyana is adjacent to Venezuela, a country with huge oil reserves. I predict that if oil turns up in Guyana, relations with the United States might warm up considerably.

OPPOSITE --

Guyana in relation to the United States. It's not much farther away than Puerto Rico.

BELOW --

The proposed state flag of Guyana. It's a great design, but I can't imagine Delta Airlines would be too happy. (Turn the flag 90 degrees counterclockwise and you'll see what I mean.)

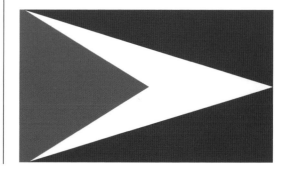

HALF-BREED TRACTS ★

Pardon the Unfortunate Name.

Before America moved west, the French were already there—trapping, trading, and surrendering. Given that there were no French girls living in the American Midwest, Franco-American men tended to marry Native American women. Sacagawea, for example, was the wife of a French trader named Toussaint Charbonneau. (Actually, she was less a wife and more a slave, but that's another story.)

As America assumed control of the Louisiana Purchase, certain lands were set aside for the native tribes, and other parcels were opened to white settlement. But none of the land was earmarked for French Indian mixed-blood peoples. By 1830, the federal government tried to rectify this problem by setting aside certain lands exclusively for this disenfranchised group.

You'd be hard pressed to come up with a more politically incorrect name than the "Half-Breed Tracts," but that's what these reservations were called. As awful as the name is, the idea was borne of a certain compassion for those of mixed race.

Unfortunately, the plan had its shortcomings. When non-half-breeds moved in illegally, the government didn't do much to stop them. Also, most homesteading laws required landowners to live on the land. But in 1834 the Half-Breed Tracts legislation repealed this provision, so many landowners began selling off their property.

Probably the most famous buyer was a guy named Joseph Smith, who gathered his Mormon followers on a Half-Breed Tract he purchased in southeast Iowa. The Latter-Day Saints then built a temple and settled in for the long haul. It's entirely possible the Mormons might have grown their sphere of influence and eventually tried to form their own state, based on their religious practices. I'm not speculating about this; that's exactly what they did. But they didn't get around to it until after they had migrated westward to Salt Lake.

Today, the Half-Breed Tracts are long gone. But perhaps some savvy half-lawyer will resurrect the legislation, find half a loophole, and voilà: a new half-casino!

OPPOSITE

Three half-breed tracts.

BELOW

The U.S. gold coin featuring Sacagawea and her baby Jean-Baptiste Charbonneau (whom William Clark nicknamed "Pomp"). Because of his mixed-race heritage, Jean-Baptiste would have qualified for land in the Half-Breed Tracts.

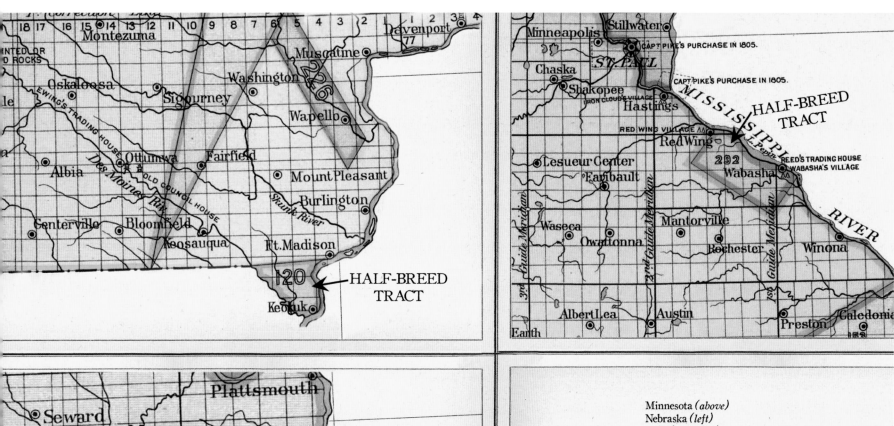

18 17 | 16 15 ⊙14 13 12 11 10 9 8 7 6 5 4 3 2 1 | 1 2 3 ⊞ 4
Montezuma

Davenport
77

Musçatine

Washington

226

Wapello

Oskaloosa
Sigourney

⊙ Fairfield

Albia Ottumwa

Mount Pleasant

Skunk River

Burlington

Centerville Bloomfield
Keosauqua

Ft. Madison

120 → HALF-BREED
TRACT

Keokuk

Minneapolis Stillwater

ST. PAUL

CAPT. PIKE'S PURCHASE IN 1805.

Chaska

CAPT. PIKE'S PURCHASE IN 1805.

Shakopee
Iron Cloud's Village Hastings

MISSISSIPPI

HALF-BREED
TRACT

Red Wing Village
Red Wing L. Pepin

Lesueur Center 292

Faribault Wabasha Reed's Trading House
Wabasha's Village

Waseca Mantorville

Owatonna Rochester Winona

RIVER

Albert Lea Austin Preston Caledonia

Earth

Plattsmouth

Seward

LINCOLN

Big Blue R.

186 Nebraska Cy.
Old Ft. Kearney

Lit. Nemaha R.

HALF-BREED
TRACT

Wilber

East

24 Beatrice Tecumseh Auburn → 155

Great Nemaha R.

Fairbury Pawnee

Falls Cy.

Minnesota *(above)*
Nebraska *(left)*
Iowa *(upper left)*

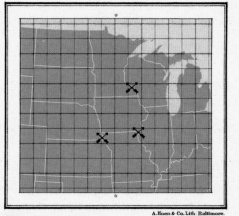

A. Hoen & Co. Lith. Baltimore.

Half-Breed Tracts

CARTE DE
LA LOUISIANE
COURS DU MISSISSIPI ET
PAIS VOISINS
Dediée à M. le Comte de Maurepas, Ministre et
Secretaire d'Etat Commandeur des Ordres
du Roy.
Par N. Bellin Ingenieur de la Marine. 1744.

ECHELLES
Lieües coûume de France de 25 au Degré.

Lieües Marines de France et d'Angleterre de 20 au Degré.

GOLPHE DU MEXIQUE

Longitude Occidentale du Meridien de Paris

★ HAZARD ★

A Presbyterian State. With No "Pope-ish" Chapels.

There's no denying that some state boundaries represent religious demarcations as much as anything else. Utah is mostly Mormon. Minnesota is the land o' Lutherans. Even California has a state religion of sorts: the cult of skin (they tan it, inject it, stretch it, and reveal way too much of it).

But what about the Presbyterians? Shouldn't they have a land to call their own?

Sam Hazard thought so.

A prominent Philadelphia merchant, Hazard developed a plan in the 1750s to colonize the then-unsettled lands of the Midwest. Planning new colonies was kind of a recreational sport in that era, much as fantasy football leagues are today.

But Hazard did more than fantasize. He recruited. By 1755 he told the general assembly of Connecticut that he had 3,500 settlers ready to go—and could add 10,000 more.

Hazard didn't overtly state that his new colony would be composed entirely of Presbyterians, but it was clear to anyone who read between the lines. His proposed charter eliminated pretty much everyone who wasn't Presbyterian. He noted, for example, that the colony would not have any "Mass Houses or Pope-ish chapels." (That's code for "Catholics: Keep out.")

Amazingly, Hazard persuaded the Connecticut assembly to relinquish its claim on lands west of the Appalachian Mountains. All Hazard needed then was the approval of the British crown.

Backchannel communications with people of influence in England met with a positive response. In the summer of 1758, Hazard made plans to travel to England to request final approval from the king. A new colony was about to be born.

But then the unexpected happened: Sam Hazard died.

The whole plan unraveled. Hazard's son Ebenezer tried to pick up the ball, but he didn't have his father's standing with the king. He then hatched a plan to bypass the king and just buy land directly from Connecticut, but the Connecticut general assembly rejected the offer.

Sam Hazard's plan had died with him.

OPPOSITE

Sam Hazard's original plan, superimposed on a French map because, at the time, the French dominated the Mississippi valley.

BELOW

After his father died, Ebenezer Hazard tried to keep the scheme alive by offering £10,000 to the Connecticut government for a swath of land that extended from western Pennsylvania to the Mississippi. Connecticut refused.

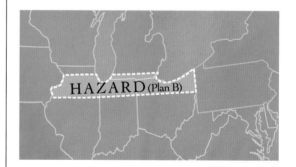

★ HOWLAND ★

Perhaps the Most Mysterious U.S. Territory.

Admittedly, Howland was never a candidate for statehood. But the story of this "lost" American island is so odd that I just had to include it.

If the name rings a bell, it's probably because of Amelia Earhart. Howland Island is the place she was supposed to land on July 2, 1937. As we all know, she never arrived—and never was found—triggering one of the great mysteries of the twentieth century.

I can't say I'm surprised she missed it. The island is a mere 300 acres—about the size of a small Midwestern farm—and just big enough for a couple landing strips.

But Howland's story was unusual even before the Earhart legend.

Because the island is located halfway between Hawaii and Australia, many believed it would be an ideal place to refuel planes making the journey between the two landmasses. So the United States tried to colonize Howland in 1935. And by "colonize," I mean the Department of Commerce dropped off a few young men (and a couple tons of canned food) and promised to resupply them after a hundred days or so.

The settlers spent their time clearing a landing strip and constructing a few ramshackle buildings. After a few months, the men were rotated off the island and replaced with fresh recruits. This exchange continued for many years. Most colonists were glad to return to civilization when their tour ended, but one, James Kamakaiwi, stayed three years. I'm not sure why he liked the tiny, barren island so much, but you have to admit it probably offered a stress-free Gilligan-like life (minus Ginger).

In 1941, things went very wrong. On December 8 (the day after the attack on Pearl Harbor), the four islanders woke up to Japanese bombers pummeling the landing strips. Remember, Howland is flat and tiny; there was nowhere to hide. Two of the colonists, Dicky Kanani Whaley and Joseph Kealoha, were killed.

If that weren't bad enough, a couple days later, Japanese subs blasted what was left of the island's meager buildings. Bombers continued to shell the runways. Eight weeks later, the two surviving colonists were evacuated. It must have been an incredibly scary two months, all alone without defense.

Howland Island would never again see any attempt at colonization—at least not by people. However, by the 1970s, the island had become overrun with cats. In the 1980s, the U.S. government removed the feline squatters.

Gradually, Howland has returned to its natural state. Today it's home to a variety of birds, dropping fresh guano on the crumbling landing strips.

Those landing strips remain the island's strange irony. The United States made great efforts to build them, the Japanese were intent on bombing them, and Amelia Earhart likely died trying to find them. Yet there is no record of any plane ever landing on Howland Island.

OPPOSITE
Howland Island—halfway between Hawaii and Australia.

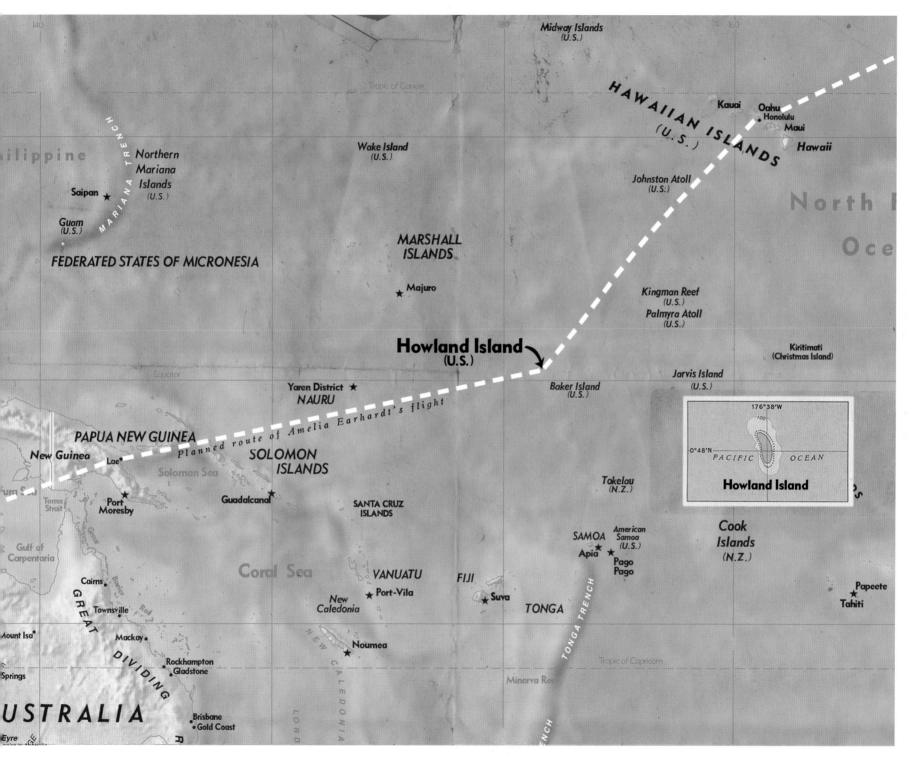

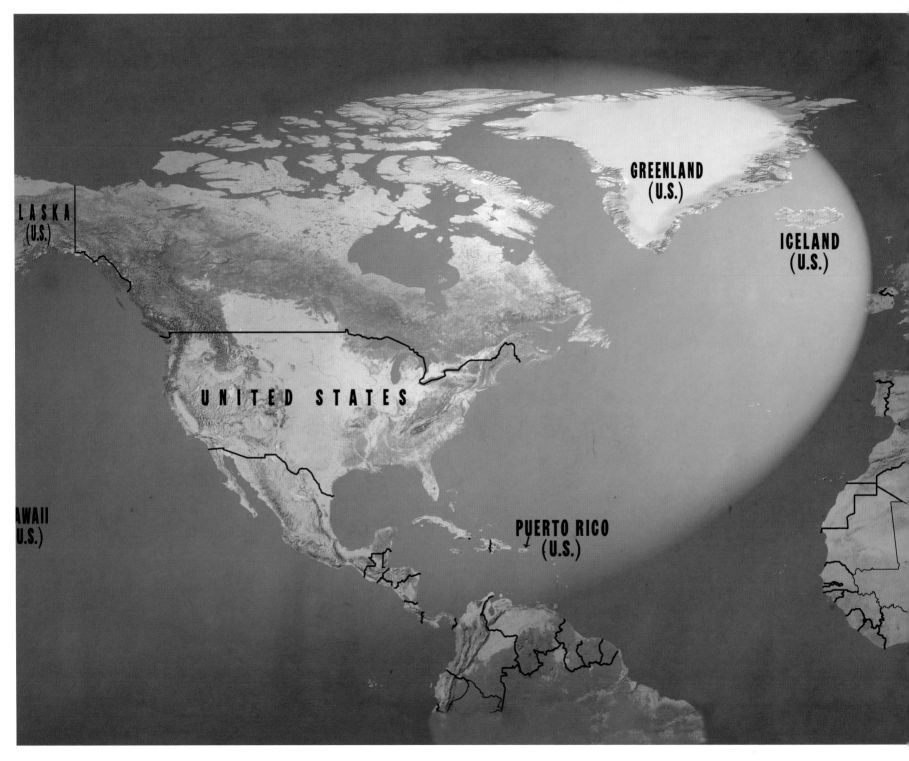

★ ICELAND ★

Strategically Important. And It Balances Out Hawaii.

Right after the defeat of Germany in World War II, the U.S. Congress considered a proposal to make Iceland the forty-ninth state. The reason was mostly strategic: Iceland would certainly be a handy place to grab coffee on the way to war with the Soviets.

Plus, America had just expended a great deal of energy trying combat the German U-boats in the North Atlantic. In retrospect, it's understandable that the U.S. military would seek a base in the region.

Enter Representative Bud Gearhart of California. He not only proposed statehood for Iceland, he also wanted to purchase dozens of other islands off America's Atlantic and Pacific coasts—and the Caribbean, too. Gearhart and his congressional cohorts figured they could just write a check for Bermuda, the Bahamas, Greenland, and anything else in the area that struck their fancy.

Their hearts were in the right place; they only wanted to keep America safe from potential foreign hostilities. But as Americans have learned, taking control of another country to accomplish that goal is mighty tricky business.

Yet, Gearhart's buyout strategy may have some merit. To wit: What if America had offered to buy, rather than invade, Iraq? I know the idea seems silly, until you run the numbers. Given the best estimates of the cost of the war, the United States could have offered each Iraqi citizen about $103,000 if they would agree to become an American. A family of five would get a cool half million dollars. Those numbers are the actual dollar costs, per Iraqi, of the war. I'll just let that sink in for a minute.

Okay, back to Iceland.

A lot of Americans supported statehood for Iceland. The nation has a long tradition of democracy, and Icelanders are well educated and quite resourceful. And you have to admit, Iceland adds a nice symmetry to a U.S. map, adding a little something in the upper right corner.

Once things settled down after WWII, support for Gearhart's proposals quickly evaporated as Americans lost interest in engaging in more foreign entanglements. Yet you could argue that Gearhardt was in fact a brilliant and visionary strategist way ahead of his time. That's because, less than two decades later, the Soviets did indeed set up a threatening base of operations on an offshore island: Cuba. For a time, it seemed Soviet nuclear missiles might just annihilate the United States.

Maybe owning some buffer real estate wasn't such a bad idea.

OPPOSITE

How the U.S. map might have looked like if Iceland had been added. The island is no farther from the American mainland than Alaska.

BELOW

Iceland is one of only two places on earth where lava is regularly produced. (The other is Hawaii.) Annexing Iceland would mean that the United States would corner the lava-rock market. That may not seem like much, but without lava-rock briquettes, gas grills would be useless. What would suburban males do on summer weekends?

★ JACINTO ★

A Better Name for East Texas. Remember Jacinto!

Over the years, dozens of proposals have been cooked up to slice Texas into multiple states. One of my favorites involves creating a new state called Jacinto.

Introduced into the U.S. Senate in 1860, this particular bill was designed to split off eastern Texas. Despite serious discussion, the bill never reached a vote.

As with nearly all Texas-splitting scenarios, this one recognized a basic truth about the Lone Star state: the eastern and western portions are markedly different. And the Brazos River is as good a dividing line as any.

But the angle about this particular plan that I find most noteworthy is the proposed name: Jacinto.

It refers to the famous battle of San Jacinto, in which Texas declared its independence from Mexico. The Alamo might be more famous, but San Jacinto was more significant. Many historians believe it was one of the more important battles of all time because it led directly to American rule of California, Arizona, New Mexico, Utah, and Nevada. Without a victory at San Jacinto, those states might have ended up in Mexico.

The battle illustrates a classic military tactic: Attack when your enemy is napping. And I mean literally taking a nap. Although the Texans, under Sam Houston, were outnumbered by the Mexican army, Houston devised a plan to sneak into the Mexican camp during the afternoon siesta. I find it hard to imagine that you could tiptoe up to a thousand napping soldiers in broad daylight, but I guess the Mexican soldiers were heavy sleepers.

Houston's forces defeated the Mexican army in eighteen minutes.

Even though the Battle of San Jacinto was never honored with a state name, at least Sam Houston got a big city named after him.

OPPOSITE

Jacinto as proposed, superimposed over a map from the period.

BELOW

Although this junction between the states of Texas and Jacinto never existed, it points to one reason that Texas never divided itself: too much rivalry over the name Texas. Nobody wanted to be East Texas or South Texas; every slice wanted to keep plain ol' Texas. Jacinto may be a name that's rich in history, but it's just not the same.

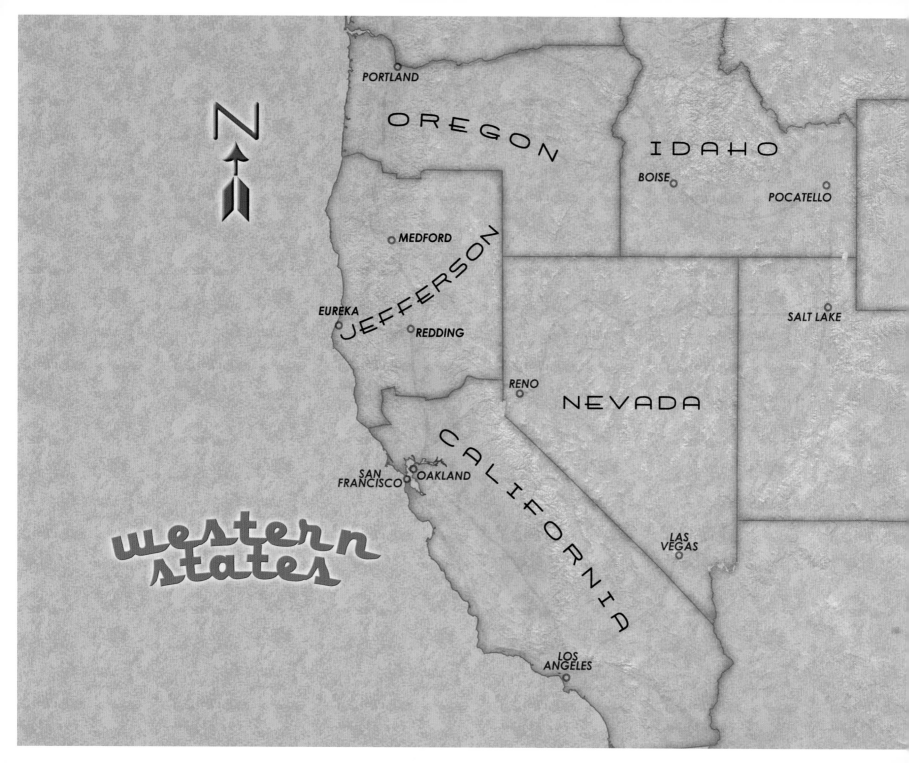

★ JEFFERSON ★

The Almost Forty-Ninth State and the War That Defeated It.

In 1941, Gilbert Gable had an idea. Why not split off the northern part of California and the southern part of Oregon to form a new state, called Jefferson?

Of course, anyone can get a wild idea, but as mayor of Port Orford, Oregon, Gable had a certain amount of influence. Before long, several border counties had endorsed his plan, and the movement began to gather momentum.

Never mind that what Gable really wanted was a better highway system. Jeffersonians (if I may call them that) were wearied by the muddy roads crisscrossing the region. They wanted the fancy new asphalt roads that the big cities were getting. Since the legislatures in Sacramento and Salem weren't overly interested in improving rural roads, the idea of Jefferson got a foothold.

By November, the newly formed Jefferson militia became involved. By "militia," I mean a couple guys with guns stopping traffic near Yreka. (Yes, that's Yreka, not Eureka. Please don't confuse these two Jefferson cities.) This "army" didn't really threaten anyone, as far as we know. But they did hand out a Proclamation of Independence—and they promised to continue seceding every week if necessary.

Of course, no one outside Jefferson gave much credence to the whole idea—at least not at first. In time, however, the notion of Jefferson as the forty-ninth state began to gain serious support across the United States.

On December 4, 1941, Jefferson unilaterally declared its independence and inaugurated its first governor. The gesture was mostly symbolic since the U.S. Congress had yet to weigh in.

Students of history will recognize how this date presented a huge hurdle to the Jeffersonians. December 4 was just a couple of days before the attack on Pearl Harbor. Once that event happened, Americans lost interest in new state proposals. World War II seemed a bit more important.

ABOVE

An imaginary poster for the forty-ninth state.

OPPOSITE

In its largest iteration, Jefferson took a big slice out of both California and Oregon.

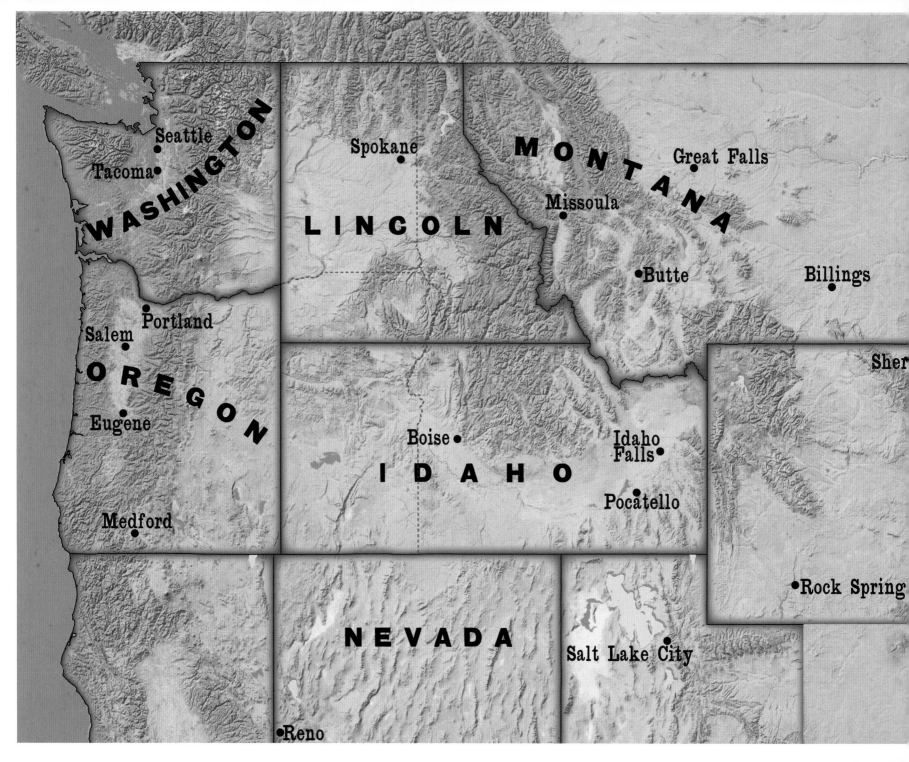

★ LINCOLN ★

A Missed Opportunity for a Better Pacific Northwest.

Idaho makes no sense. I speak from experience, having lived in the state for twenty years. Unless you're willing to navigate a treacherous mountain pass, you can't even drive from the north to the south without leaving the state.

Because it was cobbled together from leftovers, Idaho contains regions with nothing in common. They go together about as well as peanut butter and jellyfish.

The inhabitants of Idaho's desert southeast—who are mostly Mormons—look to Utah for all things spiritual and retail. The northerners, green and granola fed, really want to hook up with Spokane, Washington. In between is Boise, which is the closest thing to a city-state that you'll find in America. This problem was understood as far back as 1886, when Congress passed a bill that would have eliminated Idaho altogether, giving the south to Nevada and the north to Washington. Predictably, the Idaho legislature went apoplectic when they heard about this scheme. To escape their wrath, President Grover Cleveland vetoed the measure.

In 1907, the best plan of all was proposed, creating the state of Lincoln from northern Idaho and eastern Washington. (See map, opposite.) This brilliant idea not only solved the Idaho problem, but it also made more sense of Washington and Oregon, because both these states are profoundly divided by the Cascade Mountains. The plan ultimately failed, however: Neither Washington nor Oregon was about to give up any territory.

Ten years later, the Idaho House of Representatives once again voted in favor of the Lincoln idea. This time the plan proposed splitting only Idaho into two sections. But now it was the Idaho senate who wasn't interested.

And so the strangeness of Idaho's borders persists.

In the 1990s, ten state senators from eastern Washington threatened secession; pretty much the same drill occurred in 2005. The quest continues.

It's worth noting that the proposed name of Lincoln hasn't changed despite dozens of plans spanning more than a century. I'm not sure why. It's not like Honest Abe ever felled any trees in the region. Besides, a state named Lincoln would cause a big license plate problem. All the Illinois plates that say "Land of Lincoln" would be really confusing if there was another land of Lincoln.

OPPOSITE

The 1907 plan is a much better way to carve up the Northwest.

BELOW

The 1917 plan would have simply split Idaho in two. It solves Idaho's bipolar disorder but creates a new state that's not really big enough to be viable.

★ LONG ISLAND ★

Tired of Bosses. And Another Scary "Ism."

Long Island is, well, a long island. It's pretty big. And given all the shrimpy states in the American Northeast, it's entirely reasonable to expect that, at some point, Long Islanders would consider statehood.

The idea was seriously discussed in 1896, when The New York Times ran an article about the possibility. The newspaper interviewed sugar magnate Adolph Mollenhauer, who offered one of the greatest quotes ever to justify statehood: "We're tired of bosses and bossism."

I didn't even know bossism was a word, but it fits in nicely with some other nasty "isms," namely, fascism, Nazism, and communism. All four isms generally have the same definition: a group of bad guys who force the rest of us into submission.

And in Long Island of the 1890s, the bad guys in question were legislators from places like Buffalo, Syracuse, and Manhattan. They spent Long Island's money without looking after its interests. Or at least that's how Mollenhauer saw it.

Statehood fervor didn't ignite in the 1890s, but the idea was rekindled a hundred years later. Curiously, the issues hadn't really changed. Long Islanders complained that they funded 17 percent of the state's school budget, but got back only 12 percent.

Long Island does indeed have many of the requisites needed for statehood. Brooklyn and Queens are on the island, making its population large enough. In fact, it has more people than twenty states. And, geographically, the island is plenty large. At 1,401 square miles, it's bigger than Rhode Island and approaches the size of Delaware. Plus, Long Island has a vibrant and diversified economy.

Even though statehood has found support on the island, it's unlikely the rest of New York state would ever let it go. It may be unfair, but I guess that's how it is with bossism.

OPPOSITE

Long Island boasts a diverse set of attractions, from urban Brooklyn in the west to rural fishing spots in the east.

BELOW

The 1890s map as Aldoph Mollenhauer would have liked it.

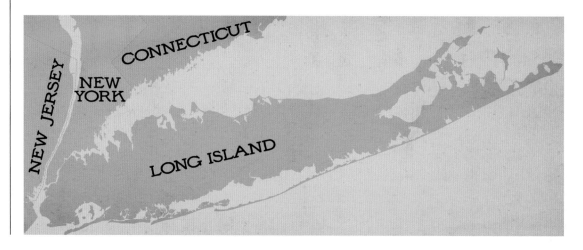

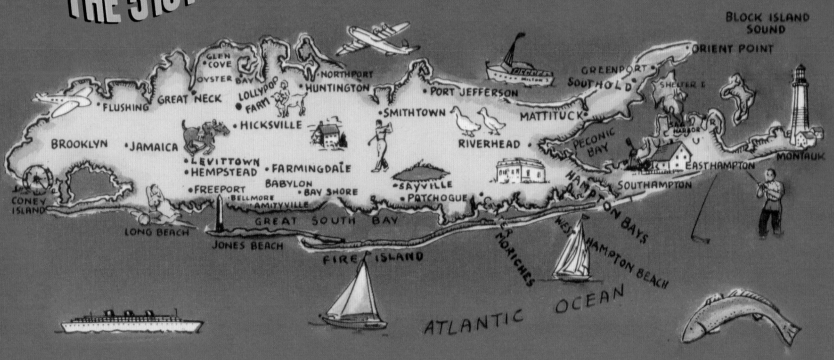

Greetings From THE 51ST STATE OF **LONG ISLAND**

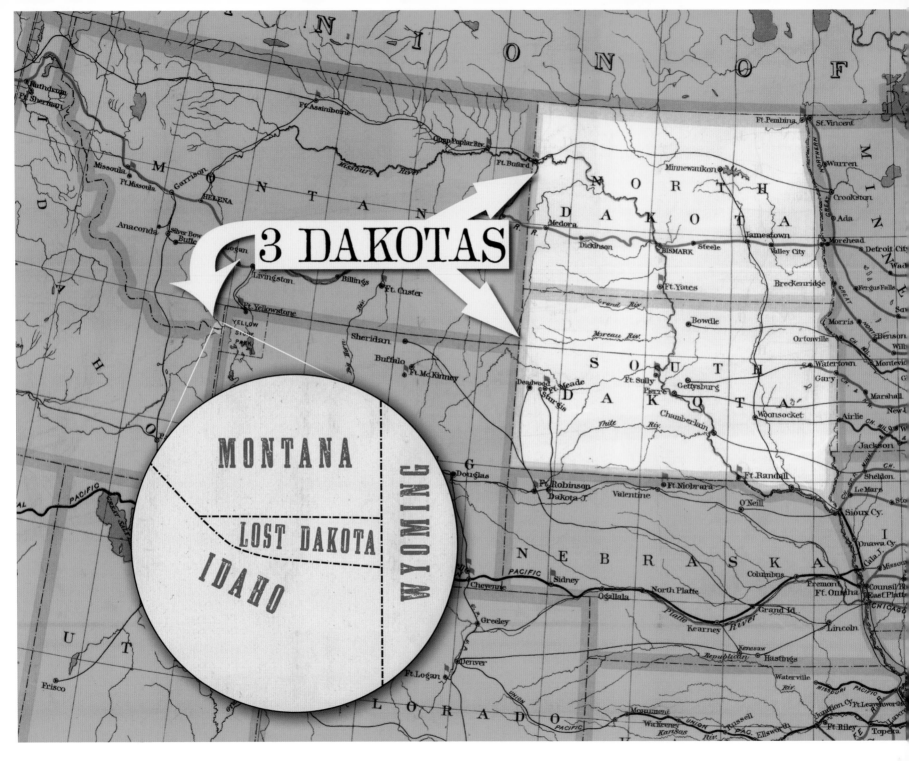

3 DAKOTAS

MONTANA

LOST DAKOTA

IDAHO

WYOMING

★ LOST DAKOTA ★

A Third Dakota. All Alone for Five Years.

How many Dakotas are there? For the longest time, there was just one huge Dakota Territory. And, today, everyone knows that the United States has two Dakotas, North and South. But for a few years in the mid-1800s, there was yet another: a tiny patch of land hundreds of miles from its bigger siblings.

Here's the story: Back when Dakota was still a territory, everyone understood that the enormous parcel needed to be sliced up into states. First Montana was carved out, then Idaho, then Wyoming. When all the chopping was done, there remained a tiny piece left over—at the point where Montana, Idaho, and Wyoming meet. Officially, this little tidbit still belonged to the Dakota Territory, even though the nearest piece of the trimmed-down Dakota was hundreds of miles away.

In retrospect, it would have been a great place for outlaws to hide out since there were no lawmen around. But in the 1860s there probably wasn't anyone living in Lost Dakota. It was only about one-third the size of Manhattan and was extremely remote. During that era, all U.S. territories expected to become states at some point, although no one proposed such a thing for this desolate patch of forest.

Even today, the land remains inaccessible. There's no road, not even a decent path. However, it does have its share of bears, not to mention the occasional buffalo herd. Yes, even today, buffalo roam the region, which is part of the Yellowstone ecosystem.

I'll admit, this little geographic hiccup isn't all that significant, but the name "Lost Dakota" is rather evocative. The anomaly lasted only a few years. In 1873, the patch was attached to Montana and became part of Gallatin County.

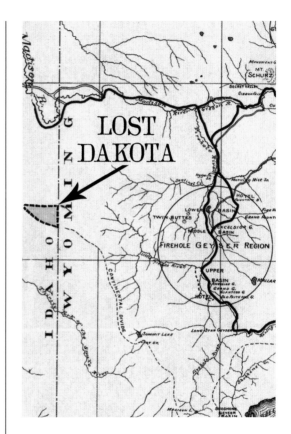

ABOVE ----------------------------

Lost Dakota superimposed over an old Yellowstone map.

OPPOSITE ----------------------------

The three Dakotas.

★ LOWER CALIFORNIA ★

How Much Would It Cost to Buy Mexico?

James Gadsden's shopping trip was perhaps the strangest in American history. In 1853, President James Buchanan sent him to Mexico to buy . . . Mexico. That's right, Gadsden was supposed to buy the country. Not all of it. More like something close to half.

Not surprisingly, Mexico wasn't selling.

Gadsden's fallback position was to purchase a small sliver, which we now call, appropriately, the Gadsden Purchase. Ostensibly, America needed flat land to run a coast-to-coast railroad, and the area just south of the border was best suited for the purpose. Since Gadsden did buy the land necessary for the railroad, you'd think his mission would have been seen as a success.

It wasn't. Gadsden's superiors in the Buchanan administration wanted more. When pressed, Gadsden suggested that the best way to acquire additional Mexican territory was to send in the troops and take over the nation.

But that would be much too messy. Gadsden left the administration, and the secretary of state then turned to John Forsyth. Forsyth was sent to Mexico to buy a substantial three-province parcel, as illustrated in the map on the opposite page.

Forsyth offered $12 million. The Mexicans declined. Forsyth then said, "Give us what we ask for . . . or we will take it." There's no record of the precise words of the Mexican reply.

Negotiations stretched throughout 1859, but the Mexican government held firm and kept its land. Had America upped its offer, however, it's possible that Mexico would have agreed to sell some additional land, especially lower California. This peninsula south of San Diego was difficult for Mexico to defend and therefore potentially expendable. The provinces of Sonora and Chihuahua—closer to the Mexican heartland—were less likely to be relinquished.

My map imagines each of these Mexican provinces as a separate U.S. state, but of course we don't know how Congress would have parsed the landscape had Forsyth succeeded. I do know that it would be great to have a state named Chihuahua. Had the state ever acquired a pro sports franchise, I think it's safe to assume the team would have been dubbed the Chihuahua Chihuahuas.

OPPOSITE

The possible outcome had Forsyth succeeded.

BELOW

James Gadsden. Are you really surprised Mexico didn't sell out to this guy?

CALIFORNIA

ARIZONA

Border prior to Gadsden's Purchase

★ Phoenix

NEW MEXICO

LOWER CALIFORNIA

SONORA

Tucson ★

Southern limit of Gadsden's Purchase

CHIHUAHUA

TEXAS

Forsyth's proposed border

MEXICO

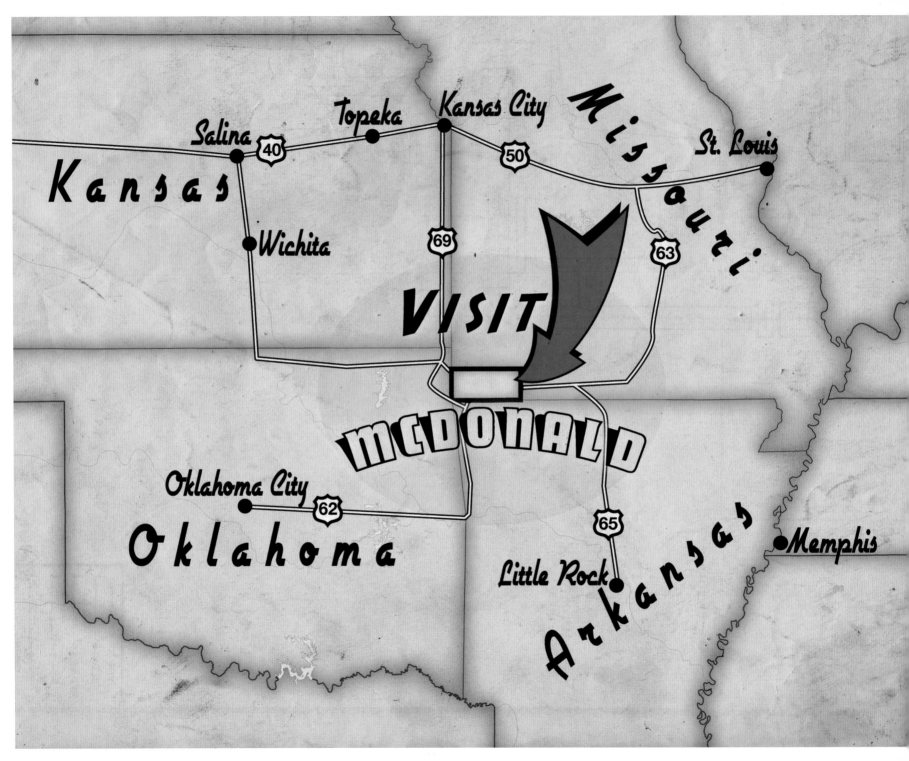

A Publicity Stunt. Or Was It?

The McDonald story has all the trappings of a great publicity stunt. In 1961, the city of Noel—in McDonald County, Missouri—was accidentally left off the state's travel map. This was a big deal back in the days before the Internet, especially for a resort town like Noel, which depended on tourism for economic survival.

So county residents decided to draw attention to their plight . . . by seceding.

If this were just your standard-issue publicity stunt, McDonald might have sent out a press release and maybe called a TV station or two. But residents took the battle to the next level. They held an election and created a provisional government. They even printed their own stamps (now highly valued by collectors).

A militia was formed (!), and soldiers were dispatched to the border to set up a checkpoint. Non-locals were given entry visas and, more important, freshly printed tourism information.

Apparently, McDonald officials soon realized they were not large enough to form their own state, so they decided to do the next best thing: annex themselves to a neighboring jurisdiction. They opened the competition to all bidders.

Arkansas governor Orval Faubus claimed to be interested in adding the county to his state. Yet it's hard to know if he was serious because, well, it's hard to take anything seriously from a guy named Orval Faubus.

The county got what seemed to be a better offer from Cherokee chief Henry Suagee, who said he was "looking for new tribe territory." Missouri governor John Dalton then weighed in, saying, "I sure would hate to lose McDonald County. There are some fine people down there."

Dalton got some welcome backup when fellow Missourians from Jasper County threatened to invade and occupy McDonald if the renegade county didn't rejoin its home state. Jasper's leaders announced that they would "suppress the rebellion and take the leaders of the insurrection into custody."

Of course, the Jasper County threat was all in good fun. In fact, the whole state-of-McDonald kerfuffle was akin to those Christmas Eve broadcasts that report that a sleigh and reindeer have been detected on radar. It's an amusing diversion, but not to be taken seriously . . . Or is it?

OPPOSITE

The kind of map the people of McDonald County really wanted.

BELOW

McDonald County was a popular vacation spot, thanks to the scenery highlighted in this old post card. One of the nation's biggest tourist attractions sprouted just a few miles from McDonald County: Branson, Missouri. In both cases, it's clear that the people of southern Missouri have a gift for promotion.

In the Ozarks, "The Land of a Million Smiles"

Entrance to Noel, Mo., South of Joplin and Neosho on U. S. 71 near Camp Crowder

★ MINNESOTA ★

A Better Way to Slice the Plains. At Least, Some People Thought So.

In 1856, everyone understood that the Minnesota Territory was too big to become just one state. It would have to be divided in half. The only question was where to draw the line.

There was considerable support for drawing an east–west boundary, as illustrated on the map (opposite). This smaller, more southern Minnesota had advantages over the current state—because northern citizens today often feel left out of a state dominated by Minneapolis and St. Paul.

Plus, people hardy enough to survive a northern Minnesota winter deserve their own state. This is the place where car-battery companies shoot their commercials because it's the only location where you're guaranteed miserable, subfreezing weather for eleven months of the year. Maybe I'm exaggerating, but it is true that northern Minnesota is colder than Anchorage, Alaska.

An east–west boundary also would have made the northern state, Dakota, really big. (And the remaining scraps of the northern Plains would have been too small to justify statehood. So the fictional West Dakota that I put on the map is rather unlikely.)

Advocates for the east–west line eventually lost, and Minnesota was split along a north–south boundary. This outcome illustrates a major theme in the design of U.S. state boundaries: Congress made a point to strategically link population hubs to less-populated areas. For example, Illinois' boundaries were drawn to connect semi-southerners to Chicago, and the Dakotas are horizontal so that the western grasslands are joined to larger cities like Fargo and Sioux City.

OPPOSITE

An 1856 proposal to split Minnesota along an east–west line. It very nearly became law.

BELOW

Any discussion of Minnesota's geography must mention the pimple at the top of the state, which is called the Northwest Angle. It's the only part of the United States (excluding Alaska) that lies above the 49th parallel. How it came to be is a complex story, but the bottom line is that 18th-century maps were often hopelessly inaccurate.

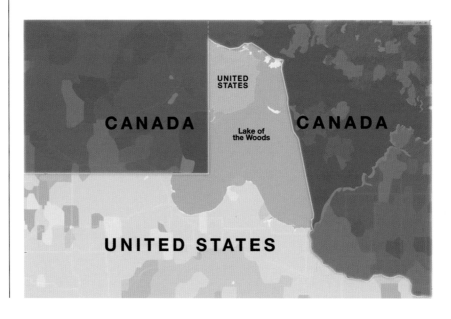

CANADA

CANADA

UNITED STATES

Lake of the Woods

UNITED STATES

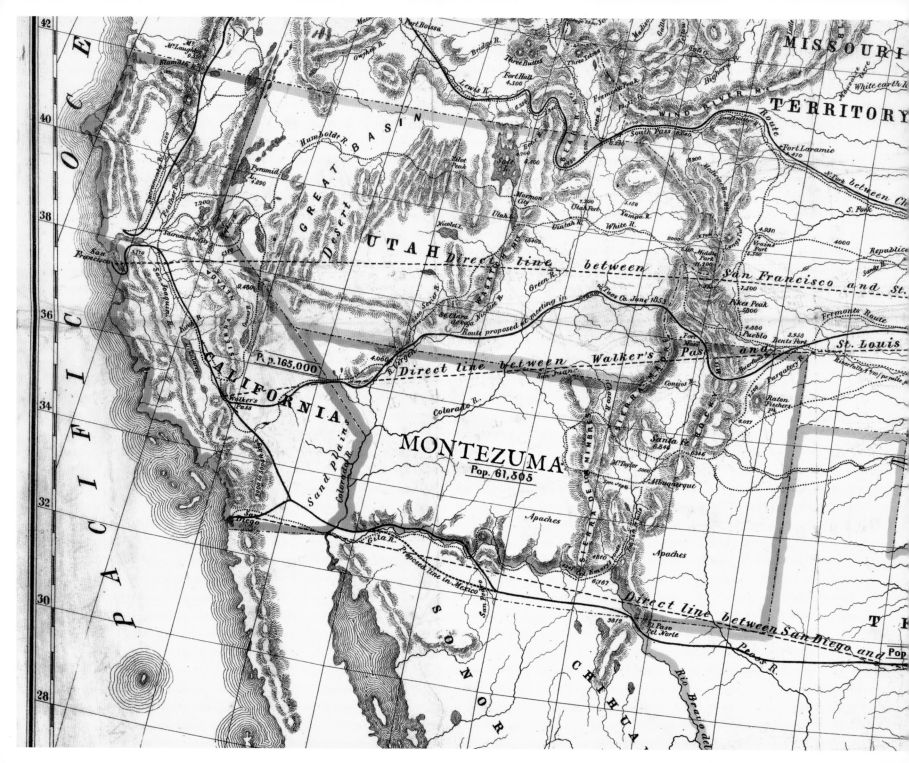

A Two-for-One Sale. And an Ill-Advised Handshake.

Today, Phoenix is the fifth-largest U.S. city, but at the beginning of the twentieth century, it had just five thousand residents. This tiny population illustrated an ongoing problem for both the Arizona and New Mexico territories: Neither had enough people to be admitted as a state.

As a result, several proposals were floated to bring the two territories into the union together, as one state. The 1903 plan proposed that this superstate be named "Montezuma." Democrats in Congress voted in favor, but they couldn't quite muster enough Republican support.

This wasn't the first time the Aztec emperor's name was appropriated. In 1887, leaders in New Mexico had proposed switching their territory's name to Montezuma, figuring that it would make statehood more palatable to the rest of the nation. It didn't. They also tried the name "Lincoln," but that didn't work either.

You've got to admit that any name is better than New Mexico, which has to be the worst state name in America. I mean no disrespect; the problem is that it's confusing. Clear evidence of the confusion exists on every New Mexico license plate. The government felt compelled to add "USA" after the state's name because too many morons thought that New Mexico was in, well, Mexico.

Speaking of stupidity, I have to mention the sorry tale of Stephen Elkins, the New Mexico Territory delegate to the U.S. Congress back in 1875. That year, population issues didn't seem to matter to Congress; the Senate and House were both overwhelmingly in favor of New Mexico statehood. But just before the final vote, Elkins entered the House floor and conspicuously shook hands with congressman Julius Burrows of Michigan.

That was a huge mistake.

Burrows had been extremely critical of Southern racial policies, infuriating congressmen from states such as Georgia and Alabama.

So when the Southern delegation saw Burrows fraternizing with Elkins, they immediately changed their votes. It was just enough to kill the statehood bill. One ill-advised handshake meant that New Mexico would have to wait thirty-seven more years to join the union.

OPPOSITE
Montezuma, superimposed on an 1853 map. Not only is the population low, but notice how empty the western half (now Arizona) is.

BELOW
Phoenix in 1885.

★ MUSKOGEE ★

Real Pirates of the Caribbean. Even More Implausible Than the Movie (If That's Possible).

The big mystery about the Muskogee story is why it hasn't been made into a big budget movie starring Johnny Depp. Here's a true story of a guy who gets kicked out of the U.S. military, joins a Creek tribe, marries the chief's daughter, consolidates several Native American nations, becomes their king, rallies the native peoples against an evil empire, gets captured and thrown into a Spanish prison, escapes, takes over a British ship, becomes a pirate . . . And there's more.

Leading a ragtag force of sixty men, our hero takes over a Spanish fort. A huge Spanish force is dispatched to capture him, but they get lost. After a series of battles, he is betrayed, captured by the enemy, and dies in a castle dungeon in Havana, Cuba.

And I didn't even get to the part where he worked as a comic actor and portrait painter.

I realize this tale strains credulity, but it's the real-life story of an American named William Augustus Bowles. In the midst of his adventures, Bowles also created a new nation-state named Muskogee, in the general area around what is now Tallahassee, Florida. His state had a capital, government bureaucrats, even its own navy. But the boundaries are hard to define, so it's unclear how far his influence extended. Nonetheless, I'm sure Bowles would consider his state much bigger than my map portrays.

The United States did not recognize Bowles's statehood claim; he was never considered much more than a nuisance by the American government. He presented a bigger problem for Spain, however, since Bowles's main theater of operation—Florida—was Spanish territory at the time.

Despite his rock star charisma, Bowles never really had the resources to match his dreams. Yet even though Muskogee eventually collapsed, it could be argued that Bowles did succeed in bringing another state (Florida) into the union. The fact that the Spanish military was unable to stop this guy—for years—illustrated just how weak they were in Florida. But Andrew Jackson noticed, and he directed the U.S. Army to invade the territory in 1812. Before long, Florida was declared American soil, and the Caribbean pirates were expelled (except for a small enclave near Orlando).

OPPOSITE

Muskogee superimposed on a map from the era. I tried to include land that fell under Bowles's control at one time or another, but one could argue that Muskogee was in fact bigger—or smaller.

BELOW

William Augustus Bowles. It's likely no American has ever lived a more movie-worthy life.

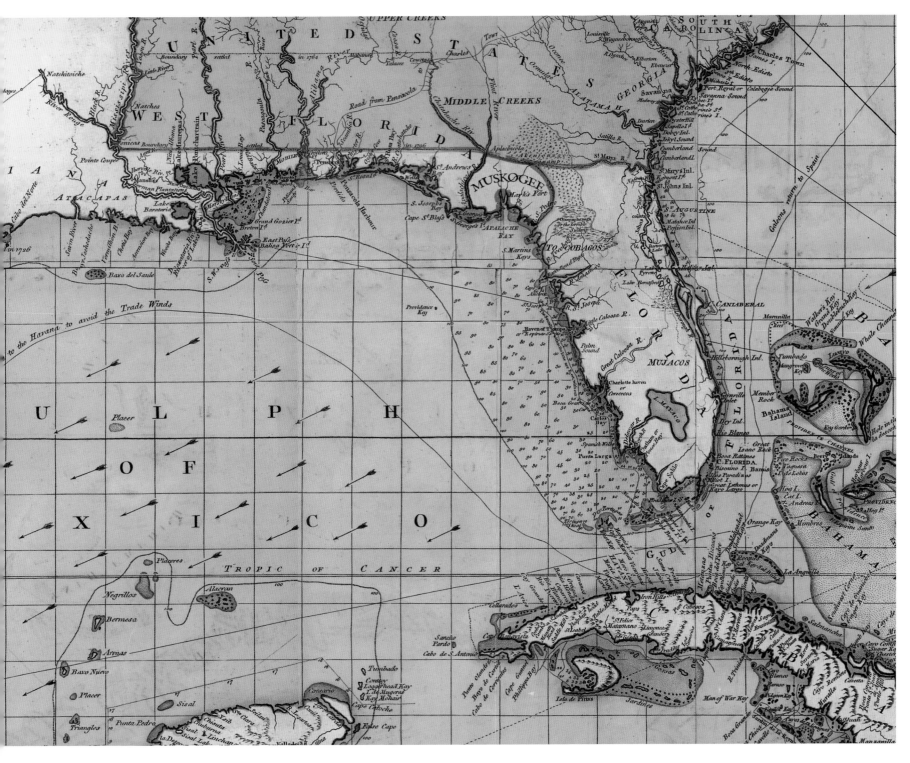

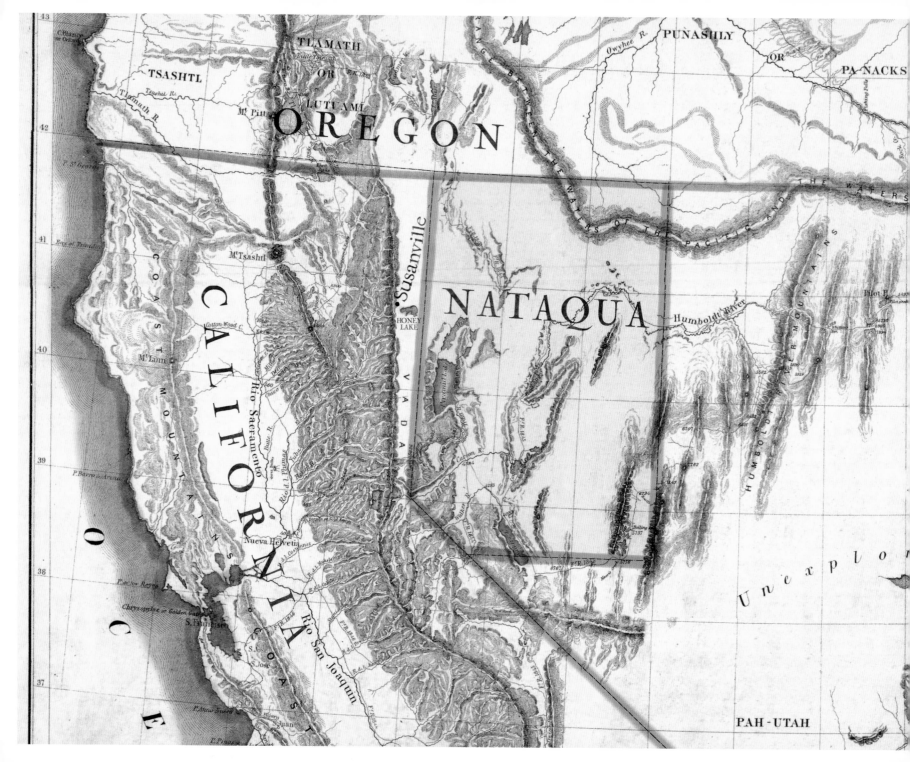

★ NATAQUA ★

The State That Was Beside Itself. Literally.

Here's something you really need if your new state is going to grow: women. The utter lack of females was the number one problem in the upstart territory of Nataqua. But I'm getting ahead of the story.

In 1853, Californian Isaac Roop stumbled upon the Honey Lake Valley, just east of the Sierra Nevada Mountains. He thought it would be the perfect place to settle. But as the first landowner in the region, Roop knew he couldn't go it alone. So he embarked on a mission to attract homesteaders to his new paradise.

Roop had no trouble convincing ex-miners to establish farms and ranches in the valley. The problem was that they were all men. Single men. They needed wives.

A desperate Roop organized dances every time a wagon train with women passed through. But apparently the valley's bachelors lacked social skills. On one occasion Roop wrote, "The boys could not say one word to them, no how."

A few months later a wagon train camped nearby. The report: "Prepared for a dance. A sad disappointment. No lady could be found. After a hard search, some were found, but (they were) engaged, so they could not come."

So when time came to create a new territory, it's not surprising that the residents of Roop's settlement chose the name "Nataqua," which is the Paiute word for "woman."

By 1858, the situation had improved somewhat. Nataqua's main community, Susanville, had grown to include about 1,000 men, 300 women, and 20,000 head of cattle.

But there was still one big problem: Nataqua wasn't in Nataqua. I'll explain.

Roop had always assumed that Honey Lake Valley was east of the California border, in an area lacking effective government. That's why he felt the need to create a new territory. But by 1858, California officials claimed that Susanville (and the whole Honey Lake Valley) was actually inside their state, and they had jurisdiction. Roop refused to admit this and fought California's oversight of the region.

Then, when Native Americans began challenging Roop's settlement, he decided that he really was in California after all, and maybe the state could send some help. When the tribal issues were resolved, Roop again argued that Susanville was not in California. How convenient.

Eventually, somebody conducted a survey, which revealed that Susanville was squarely in California. Nataqua then evaporated. Nevada (minus Susanville) came a bit later.

Today, they seem to have plenty of women.

OPPOSITE

Nataqua superimposed on a map from the era. Look closely and you'll see Susanville and Honey Lake outside the borders of Nataqua. Oops.

BELOW

Isaac Roop's original cabin survives.

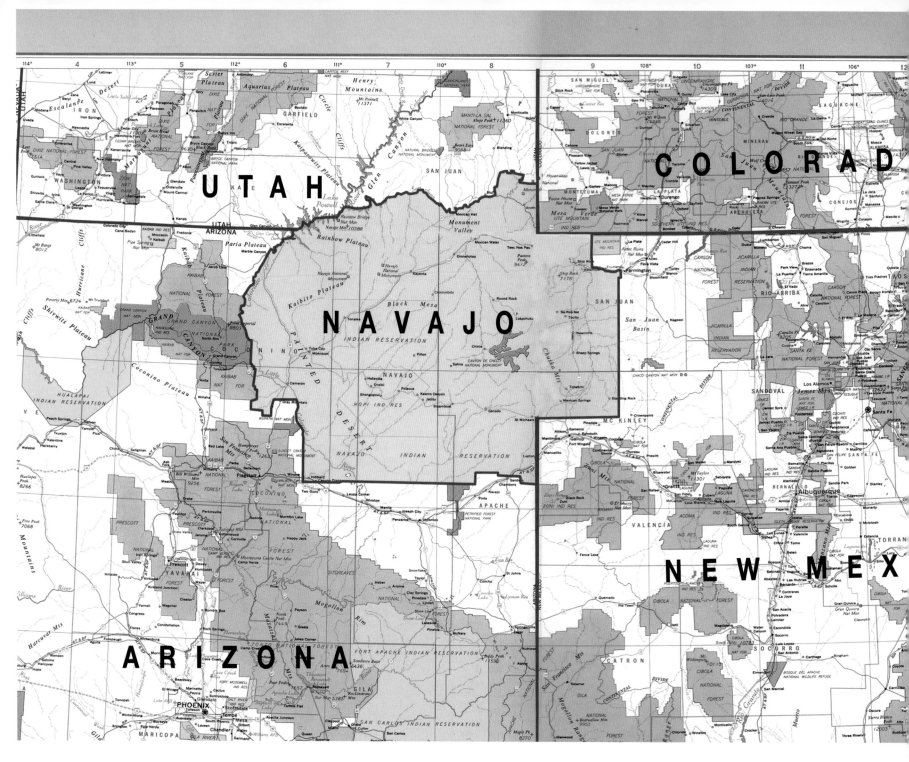

★ NAVAJO ★

It's Big Enough. And It's Got That Valley.

Everyone agrees that Europeans pillaged the homeland of the first Americans. But there's not much we can do about it today, is there? What's done is done, right?

Not so fast. Back in the 1970s, Navajo tribal chairman Peter McDonald offered a novel idea: make the Navajo homeland the fifty-first state. His belief was that only through statehood could his people achieve true equality with non-Indians.

His proposal was logical. Current U.S. reservation law tends to put native tribes in a nebulous position. They have some autonomy and self-rule, but on significant matters the tribes tend to lose out to their non-native counterparts.

One example, among many: Nearly every Native American resident living on the Ft. Hall reservation supports the building of a Vegas-style casino. But because the reservation is surrounded by Mormon-dominated Idaho, the state legislature and the courts have blocked the tribe's efforts. This tale is repeated in different iterations all across the United States.

The Navajo state idea was especially intriguing because—unlike most reservations—the Navajo homeland is huge. And it's chock-full of valuable commodities including oil, coal, and uranium.

The biggest downside I can see is that the famous Four Corners would have been zapped out of existence. (In case you missed that day in geography class, the Four Corners is the only place in the United States where four states touch each other.) It's a popular tourist spot, though I imagine business would trail off under statehood. Who would want to visit "One Corner"?

So what happened to Peter McDonald's proposal? It received national attention for a while in 1974, but it faced tremendous obstacles. Specifically, it would have required the support of three different states (New Mexico, Arizona, and Utah), and states generally don't like giving up territory. Plus, the reservation's population hasn't grown as McDonald had hoped.

Still, every so often statehood proposals crop up on Indian reservations. This one may rise again.

OPPOSITE

Map of 1970s southwestern United States, with an overlay of the proposed Navajo state.

BELOW

Iconic Monument Valley is located on the Navajo reservation.

★ NAVASSA ★

And the Guano Islands. If It Has Bird Poop, It's Ours.

Let's say you're sailing around the world, and you come upon a previously uncharted island. On closer inspection, you discover the beach is covered in bird droppings. That's good, because it means that you can then plant the flag and claim the island as a new U.S. territory.

It's the law. And it isn't some obscure, never-enforced footnote. This is a big deal called the Guano Islands Act. First passed in 1856, the act declares that any citizen can claim any island worldwide as United States land, provided that no other country has gotten there first. Oh, and it has to have a lot of seabird dung.

You see, back in the mid-1800s guano was a prized fertilizer. Many tiny ocean islands had huge supplies of the stuff. Why? A bird rests on a rocky outcropping, and while there he relieves himself. Then the next bird comes by. This goes on for thousands of years. You get the idea.

If you doubt the value of seabird doo-doo, consider a scientific study published in 1859: A particular "tired" acre of Georgia farmland grew 808 pounds of cotton. The adjacent acre received a dusting of guano, and the yield jumped to 1,800 pounds of cotton.

Smart businessmen wanted to scoop up the stuff and sell it back on the mainland, but they needed some legal cover—thus the Guano Islands Act. The side effect was that the islands claimed then became a part of the United States. The country added more than one hundred islands via the act.

Each of these islands has a story, but Navassa's is especially compelling. The two-square-mile Caribbean island had so much guano that in the late 1800s a Baltimore mining company sent 140 African American workers to begin extraction. It was a major operation, complete with a railroad, warehouses, even a church.

But there was a problem. The guano wasn't easy to extract, which meant that workers had to chip away with pick-axes all day long. The operation went on for years. And it was hot. Really hot.

And what did the workers do on their day off? The tiny island wasn't even big enough to play the newly invented game of baseball. It was a miserable existence. Tensions rose until a rebellion erupted, and five supervisors were killed in the fighting. Eighteen workers were sent to Baltimore to face murder charges; three were convicted and faced execution.

Defense lawyers then took a creative stance, arguing that the island wasn't U.S. territory, making the trials invalid. But in 1890, the Supreme Court weighed in, ruling that the Guano Islands Act was indeed constitutional. The ruling still stands. As for the workers, President Benjamin Harrison commuted the death sentences.

Today, Navassa and several other nearby guano islands remain U.S. territory. Although statehood for the guano islands has never been proposed, combining them all would theoretically create a healthy-sized territory. And though most of the islands are pretty remote and desolate, you could have said the same thing about Las Vegas in 1930.

OPPOSITE --

Navassa Island, two square miles of "solid gold." If you look closely, you can see the lighthouse that still stands.

CLAIMED
BY GUANO ISLANDS ACT

NAVASSA U.S.

A.K.A. Vermont. The Flag Problem Emerges.

Vermont sure seems like it should have been one of the original thirteen colonies.

It wasn't. In fact, early Vermonters wanted nothing to do with the United States. In 1777, they declared themselves a separate country, which they named New Connecticut.

The name choice may seem a bit odd since the "old" Connecticut wasn't really all that old. But tacking on "New" was one of the era's trendy fads.

It's just one of a number of curious state-naming trends that have had their fifteen minutes of fame. Thomas Jefferson fancied adding "ippi" to the end of state names. He proposed two states with the suffix, but Congress vetoed the idea (although Mississippi did eventually enter the union).

Then there was the popular trend of naming states after Native American words. Unfortunately, tribal languages were often massacred in the process. For example, the Meskousing River was mistakenly written as Ouisconsing and, later, Wisconsin. That same river's name was sometimes transcribed as Ouiscon-sint, which is the origin of Oregon. That's right, both Oregon and Wisconsin are named after the Meskousing River. Go figure.

But let's get back to New Connecticut. After six months, citizens agreed to change the name to Vermont, but for more than a decade they remained independent from the United States. New York kept trying to swallow up the territory, as did New Hampshire, but plucky Vermonters managed to join the union—on their own terms—in 1791.

All was not perfect, however. Statehood caused an immediate dilemma. As the fourteenth state, Vermont (along with Kentucky, number fifteen) forced the design of a new American flag, because the existing design had one star and one stripe for each state. The obvious solution was to revise the flag to include fifteen stars and fifteen stripes. That solution worked fine for a while, but it didn't take a genius to see that a growing nation would result in an increasingly cluttered flag. Eventually a better design was implemented: New states would have their own star, but the flag was put on a stripe diet, slimming down to thirteen stripes (for the original colonies), which it's had ever since.

OPPOSITE

New Connecticut, shown on a map from the era.

BELOW

The 15-stripe flag. Later, the design was trimmed to 13.

★ NEW SWEDEN ★

Its One Mistake: Crossing the Dutch Boys

The only taste of Swedish culture most Americans get these days is the occasional trip to IKEA. However, Swedish influence in the United States predates the monstrous blue and yellow retailer by some 350 years.

In 1638, a group of Swedes settled near what is now Wilmington, Delaware. Over the next two decades, Swedish settlements popped up all along the Delaware River. It's possible that this growing colony, called "New Sweden," might have eventually joined the United States, just as other independent republics, such as Vermont, did.

But before the colony reached a critical mass, the Swedes made a huge mistake. They tried to seize Fort Casimir, a nearby Dutch outpost. The good news: The Swedes succeeded. The bad news: This made the Dutch really angry. The next year, the Dutch government sent a massive troop surge to recapture the fort. The show of force overwhelmed the Swedes and effectively put an end to New Sweden.

But the defeat didn't deter Swedish interest in America. In the late 1800s, millions immigrated to the United States, settling first in the Minneapolis–St. Paul area and later farther west, into North Dakota and Montana. There's even a New Sweden in Idaho, although it's hard to imagine a place with less resemblance to the mother country. I once asked a New Sweden farmer why he'd left Kalmar, Sweden, in 1910 to settle in Idaho's dusty, treeless desert.

His one-word reply: "Land."

The answer summed up three hundred years of American history.

OPPOSITE

If you can't find New Sweden, it's because the era's mapmakers liked to use Latin names. Thus, New Sweden is Nova Suécia.

BELOW

The log cabin might seem like a classic American invention, but it's not. Swedes developed the technology first and brought it to America. I guess it makes sense. All those interlocking pieces of wood foreshadowed IKEA.

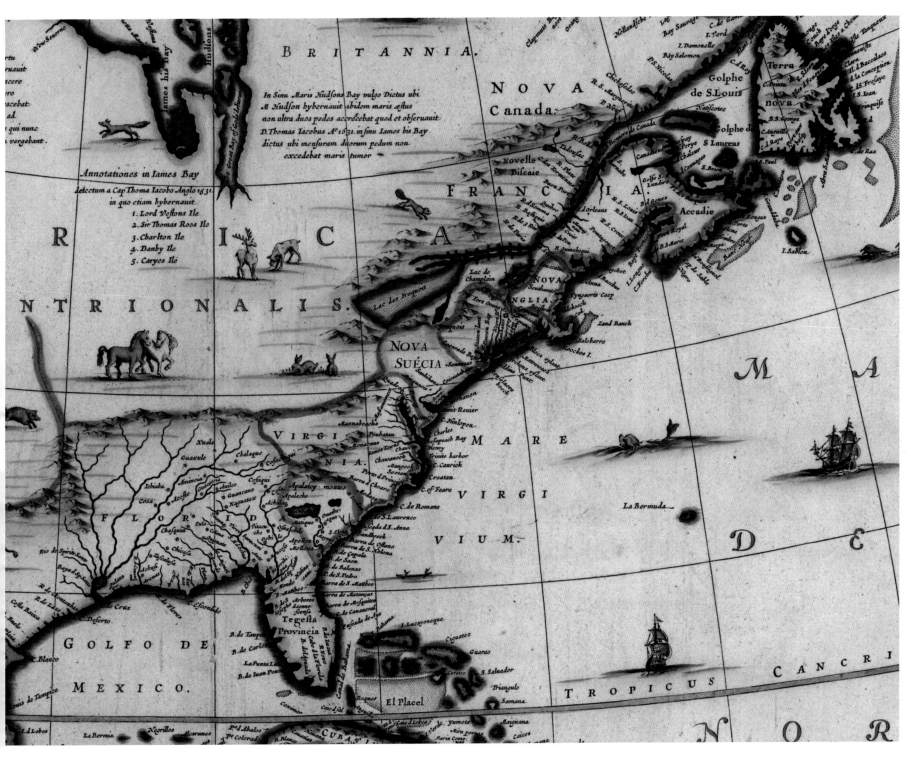

A Logical City-State. But the Name's a Problem.

New York City has a lot of people. The rest of the state does not. And therein lies the rub. All those city slickers don't want a few country bumpkins telling them what to do.

This problem has existed in New York City for more than a century. As early as 1861, the city's mayor was campaigning for statehood. The matter heated up in the 1970s, when Major John Lindsey and others took up the idea.

Their motivation was sound enough: a large percentage of state taxes collected from city residents never came back to help pay for city services. But it was really more about control. New York City thought it was big enough to handle its own affairs.

One of the intractable problems of this statehood proposal is the potential name of the new state. It can't be "New York City" since, well, it wouldn't be a city anymore. And the city doesn't have any nicknames that would be suitable. I mean, you can't really name a state "The Big Apple" or "Gotham."

So if New York City became a state, you'd have to name it New York. But then what do you call the rest of the state? On my map I settled on Empire since New York's nickname is "The Empire State." (That's how the Empire State Building got its name, of course.) But that's still not ideal. Other options, such as North New York or Upper New York, violate an unwritten rule of state names: They can't be more than two words.

OPPOSITE

Imagining New York City of the 1970s as a state, with the rest of New York named "Empire."

BELOW

Manhattan in the mid-1800s. Even back then the mayor wanted statehood.

★ NEWFOUNDLAND ★

Newfoundlanders Were in Favor. And So Was Chicago.

Most Americans assume that Newfoundland was always a part of Canada, but that's incorrect. As recently as the 1930s, Newfoundland was an independent country.

When it became clear that Newfoundland would be better served by joining a larger nation, most Newfoundlanders wanted to attach themselves to the United States. Indeed, polls taken in 1947 report that 80 percent of the population wanted to become Americans. So why didn't Newfoundland become the forty-ninth state?

The problem lay with the U.S. government, which had absolutely no interest in annexing Newfoundland. Had the United States made even the most modest of overtures, it's quite possible that Newfoundland would now be part of the union.

It seems that the only Americans openly enthusiastic about Newfoundland statehood were the editors of the Chicago Tribune; the paper was a breathless supporter of the idea. This was back in an era when the personal crusades of newspaper owners could influence history. William Randolph Hearst, for example, famously manufactured facts to persuade America to go to war with Spain. Nevertheless, *Tribune* owner Robert McCormick wasn't quite as successful at mobilizing public opinion regarding Newfoundland.

By 1949, Newfoundland had agreed to merge with Canada.

It may have been a huge mistake.

Many believe that the Canadian government's policies messed up Newfoundland's fishing industry, and—with the economy decimated—young Newfoundlanders began leaving in droves. Things spiraled downward.

Talk of Newfoundland statehood resurfaces every so often, especially whenever Quebec threatens to leave Canada. Given Newfoundland's distinct history and isolated geography; it remains among the likeliest provinces to extricate itself from Canada. If that happens, the United States is certainly a probable place for it to land.

OPPOSITE

The province of Newfoundland includes both the island of Newfoundland and the mainland region, known as Labrador.

BELOW

The Newfoundland dog breed is known for its sweet disposition and water-rescue ability. But don't call them "Newfies": Newfoundlanders consider the term pejorative (for the dogs and the people). But it's okay to refer to a Labrador as a "Lab."

ES

Ke

Southampton
Island

Coats Island

Mansel Island

Hudson Strait

LABRADOR SEA

HUDSON BAY

Churchill

hill

Nelson

OBA

Severn

Belcher Island

Feuilles

Nairn

N E W F O U N D L A N D
(U.S.A.)

Goose
Bay

Gander

St John's

La Grande Rivère

Fort George

JAMES'S
BAY

*Akimiski
Island*

Newfoundland

Ile d'Anticosti

Gulf of St. Lawrence

ONTARIO

Fort Albany

Albany

C A N A D A

Q U E B E C

PRINCE
EDWARD
ISLAND

Charlottetown

NEW
BRUNSWICK

Lake Nipigon

Quebec

Halifax

St John

NOVA
SCOTIA

Thunder Bay

Lake Superior

Montreal

ATLANTIC OCEAN

U N I T E D

Duluth

Lake Michigan

Ottawa

Minneapolis

St Paul

Toronto

Lake Ontario

Boston

Mississippi

Lake Huron

Buffalo

S T A T E S

Milwaukee

Detroit

Lake Erie

New York

Chicago

Cleveland

Philadelphia

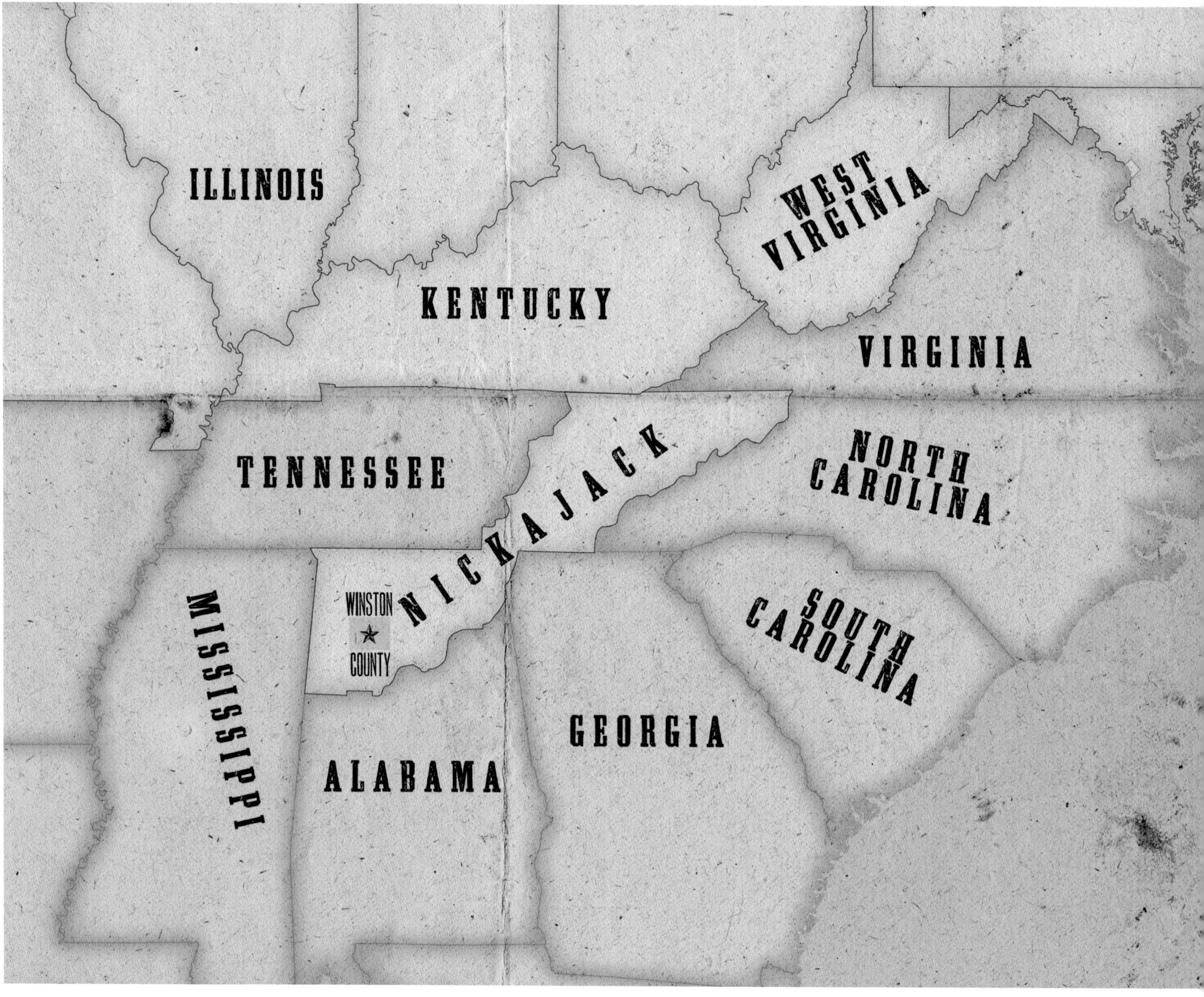
ILLINOIS
KENTUCKY
WEST VIRGINIA
VIRGINIA
TENNESSEE
NICKAJACK
NORTH CAROLINA
WINSTON
COUNTY
MISSISSIPPI
SOUTH CAROLINA
GEORGIA
ALABAMA

★ NICKAJACK ★

A Redneck Berkeley. And the Pro-Union South.

In the Civil War era, many Southerners opposed seceding from the Union. Indeed, public opinion in the South wasn't nearly as unified as your seventh-grade textbook implied. That's because slavery benefited wealthy plantation owners, not the poor dirt farmers of the Appalachians. Many of these hill people resented being dragged into what they considered to be a rich man's war.

The epicenter of anti-secessionist sentiment was Winston County, Alabama. On July 4, 1861, a group of 2,500 people gathered to declare their neutrality, and the "Free State of Winston" was born. Winston soon became a haven for anti-war types—sort of the Berkeley of the Civil War era.

Of course, the Confederates weren't very pleased, and they staged regular raids into the county to gather unwilling conscripts. Winston's draft dodgers didn't flee to the North, they just hid—and Winston County had lots of valleys and hollows to hide in. Some were caught and ended up fighting for the Confederacy; others joined the Union army; and some just stayed in hiding.

And Winston wasn't alone. People from surrounding counties shared the sentiment. In fact, all of eastern Tennessee voted against secession, as did the people of northwest Georgia.

A plan was formulated to connect these disaffected Southerners and create a neutral state called Nickajack. You can't argue with the logic: If the southern states could secede from the Union, why couldn't a few counties secede from the secession?

But what if there had been a pro-Confederacy movement within Nickajack? Would that have meant a secession from the secession of the secession? Imagine the conversation:

"Are you a secessionist?"

"Yes, I advocate secession from the secession."

"Then you are not a secessionist, sir!"

"How dare you suggest my secession from the secession is not a true secession."

"The only valid secessions are the original secession and the secession from the secession of the secession!"

Now you know why they just wore blue and gray uniforms.

The Nickajack idea won a lot of grass-roots support and had a template for success. Just to the north, a group of pro-Union counties had already separated from the South and formed a new, pro-Union state, called West Virginia. Nickajack hoped to follow the same path.

But Nickajack needed a strong leader, and none came forward. By the end of 1861, the dream of a state of Nickajack had faded into history.

OPPOSITE

Nickajack never had an official map, so this map is somewhat of a conjecture. We know that majorities in Winston and the surrounding counties opposed secession, as did Dade County, Georgia, and most of eastern Tennessee. Combining those counties gives a pretty good idea of Nickajack's boundaries. Had the process moved forward, Nickajack might have gained even more of northwest Georgia and parts of western Virginia.

★ NO MAN'S LAND ★

Land of Evil-Doers. And Cimarron Strippers.

It sounds like fiction, but there really was a sliver of the American West called "No Man's Land," beginning in 1845. Located just north of Texas, the region was not part of any state or territory. Thus, there was no law enforcement because, well, there were no laws to enforce.

As you can imagine, outlaws flocked to No Man's Land. *The New York Times* described it as a "thieves' paradise" and a refuge for horse robbers and "evil-doers." (Yes, the paper really did use the word "evil-doer" long before the George W. Bush presidency.)

Soon, the bad guys had company. In the 1860s, a group of unauthorized settlers (that is, squatters) established ranches in the region. To keep the peace, the newcomers organized vigilante committees, a system of self-policing that actually worked.

To make things official, representatives from No Man's Land asked Congress to officially recognize the territory, which by then was known as Cimarron, or the Cimarron Strip.

Congress refused. The reluctance may have been due to the territory's size. During that era, territories did eventually become states, but Cimarron seemed far too small for statehood.

The quest received its final death knell when Congress opened the adjacent Oklahoma Unassigned Territory for settlement. Since land there was much more desirable, Oklahoma attracted a large percentage of Cimarron Strippers (maybe I shouldn't call them that). Soon, almost two-thirds of Cimarron's residents had rushed to Oklahoma.

With its land emptied, Cimarron's hope for official territorial status—and statehood—ended. The Cimarron strip soon joined Oklahoma, where it became known as Beaver County.

OPPOSITE

No Man's Land, superimposed on a railroad map from the era.

BELOW

This unretouched photo illustrates the challenge of No Man's Land. High winds and a lack of rain often resulted in dust bowl conditions. So when better land became available in Oklahoma, most No Man's Land residents left for greener pastures.

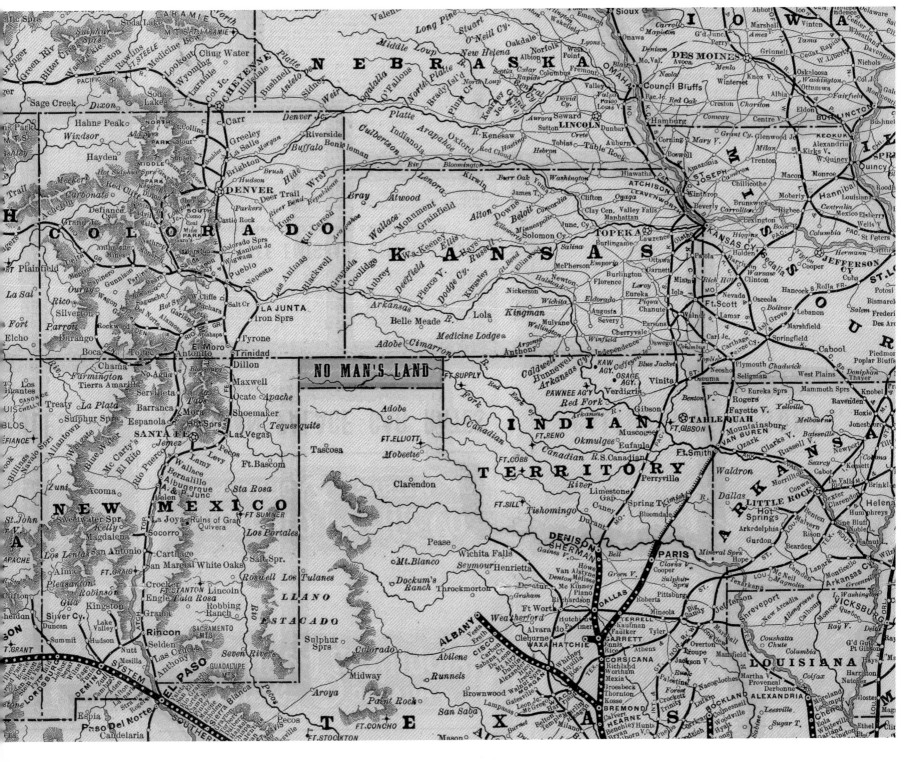

★ NORTH SLOPE ★

Not as Unlikely as It Might Seem. But It Needs a Better Name.

On the face of it, the 1992 proposal to slice off the top of Alaska and form a new state seems rather ridiculous. But one can make a surprisingly cogent argument in favor of the idea. Northern Alaska is home to a huge oil reserve, and the federal government's response has been to pump out the oil and sell it to residents in the lower forty-eight. That's just fine with the native Inupiat and Yupik peoples. They make a tidy commission on the oil operation.

But they draw the line when it comes to messing with their traditional ways of life. Which brings me to the main reason that north Alaskans have suggested secession. They were frustrated by a state supreme court decision that forced them to stop hunting and fishing in the manner to which they had grown accustomed. They've been living off the land for centuries, so you'd expect them to be a little miffed.

The Inupiat and Yupik peoples argue that their voice is not heard and that the system is rigged against them. That is, when issues affecting native peoples move through the political process, they are outnumbered by urban Alaskans in cities like Anchorage and Juneau.

Yes, I too thought that the term *urban Alas-*kans was oxymoronic. But compared to Barrow (population 5,000), Anchorage (population 250,000) is quite cosmopolitan.

The idea that small rural areas shouldn't be ruled by big cities is one of the main reasons that states exist. So it seems very American to allow the hardy souls on the North Slope to govern themselves.

If you still think statehood for northern Alaska is unworkable, looking eastward might change your mind. A remarkable precedent was set in Canada in 1999, when the national government gave the Inuit people their own territory, named Nunavut.

If the United States did something similar, the new territory probably wouldn't be called North Slope. If the native peoples have a choice, I'm betting they'd go with a more traditional name.

OPPOSITE

The borders of a new Alaskan state, as proposed in 1992.

BELOW

This upside-downish map offers a better sense of Alaska's relationship to the Arctic, Canada, Russia, and Greenland.

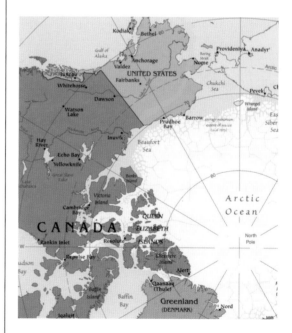

★ PANAMA ★

A Win-Win? Teddy Acquired It. Jimmy Gave It Back.

Why would America be interested in adding a new state located in the heart of the Central American rain forest? To understand this little piece of history, you have to look all the way back to 1492.

Remember Christopher Columbus's original goal? He set sail in search of a shortcut from Europe to Asia. He never did find the passage because he kept bumping into land. The North and South American continents are a big obstacle to east–west trade. Indeed, for hundreds of years, sailors dreamed of slicing the Americas in half to create a quicker shipping route.

Technology didn't make this idea feasible until the early 1900s. But by then there was the sticky problem of land ownership. The skinniest parts of the Americas were owned by other countries.

The United States wasn't about to let that minor detail get in the way of progress. So, under President Teddy Roosevelt, it supported a revolution in northern Columbia, which led to the creation of a new America-friendly country: Panama.

The Panamanians then gave the United States a slice of land—called the Canal Zone—and we dug our so-called big ditch.

But in the 1960s, things started to deteriorate. Civil unrest (that staple of the 1960s) began to grow. Panama wanted its land back, yet many Americans believed the big ditch was vitally important; they urged the federal government to do whatever was necessary to quiet the revolts and keep the canal.

And so in 1964, *Parade* magazine and other publications began to advocate statehood for the entire nation of Panama. In theory, statehood would mean stability and prosperity—a win-win for both the Panamanians and the United States. Support was thin, and the idea died officially in 1977 when President Jimmy Carter returned the canal to Panama.

OPPOSITE

This map illustrates how the state of Panama would have looked, along with other U.S. territory in the 1960s. (Panama is the green blotch near the bottom.)

BELOW

This CIA map provides a closer look at Panama. The mountains were clearly a big obstacle to canal building, except at the gap near Balboa.

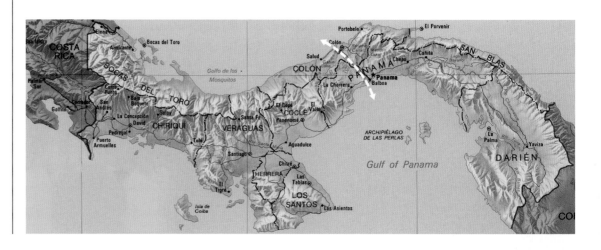

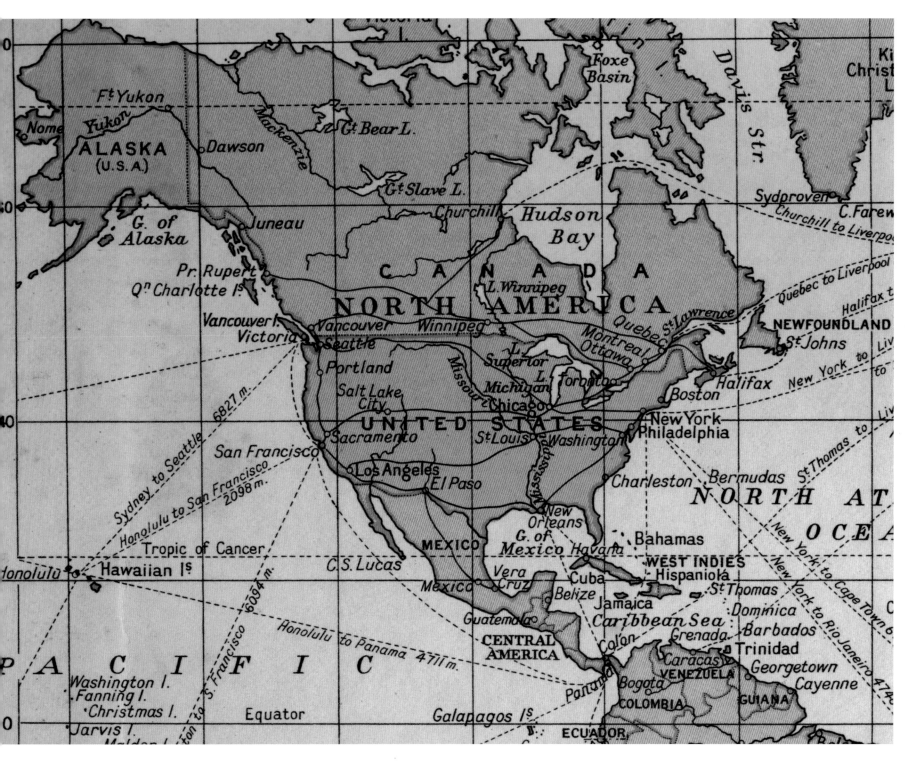

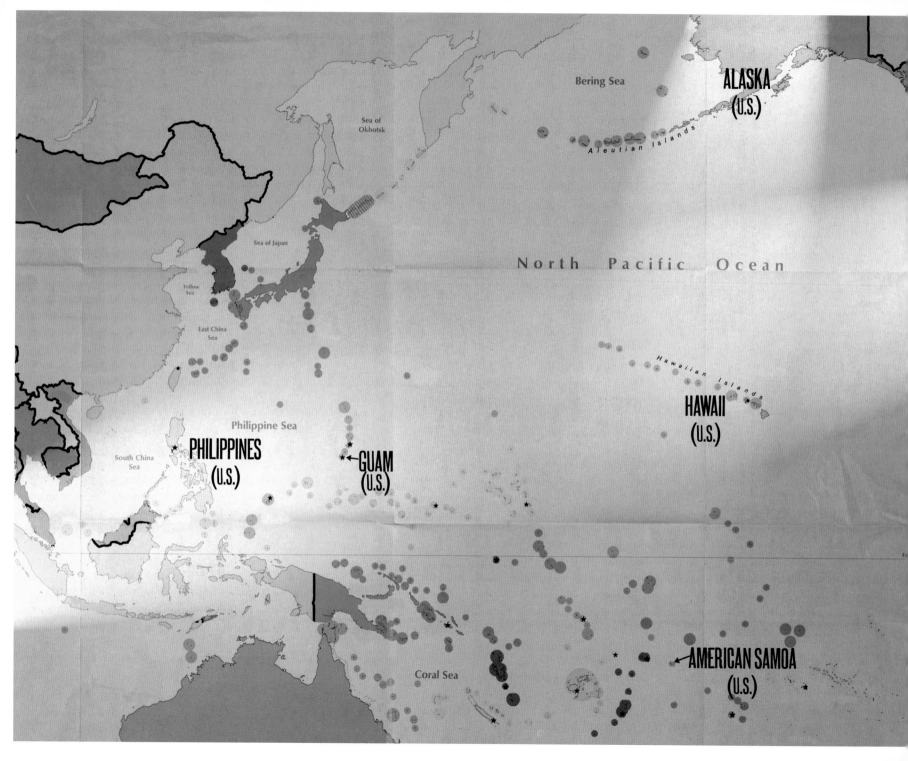

★ PHILIPPINES ★

The First Quagmire. Foreshadowing Vietnam and Iraq.

Even though I grew up in the 1960s and '70s, I have a pretty good understanding of 1940s world geography. That's because my elementary school couldn't afford new maps. (We even said the Pledge of Allegiance to a flag with forty-eight stars.)

So I was well aware of the Philippines' brush with statehood. See, on those old maps, the Philippines was identified as U.S. territory.

Here's why: The United States received the Philippines from Spain in 1898 as a result of the American victory in the Spanish-American War. Spain had held the islands for hundreds of years, so when they pulled out, a power vacuum quickly emerged. U.S. president James McKinley, like most observers, believed that if America didn't take over, insurgents and warlords would make things pretty awful for the Filipino people. Reading his speeches, I think McKinley honestly believed he was doing the right thing.

So America stayed.

At the time, it was customary for U.S. territories to eventually become states. Philippine statehood was debated at length among U.S. leaders. If the idea seems farfetched, remember the example of Hawaii. Both Hawaii and the Philippines were Pacific island groups with populations that were ethnically distinct from most mainlanders. Yet Hawaii eventually achieved statehood, and the Philippines did not. Why?

The simple answer: The Philippine islands are really, really far away.

But that isn't the full story. It's clear that racism also played a role. Many in Congress didn't want millions of Filipinos to have a voice in American affairs. (The 150,000 Hawaiians didn't seem quite so threatening.) As a result, for many years the Philippines floundered in a nebulous status. They were recognized as American territory, but by the 1920s the possibility of statehood had faded.

In 1946, the Philippines received full independence from the United States. The situation went well for a few years, until Ferdinand Marcos decided that being president wasn't enough and installed himself as dictator.

Maybe statehood wasn't such a bad idea.

OPPOSITE

This map shows major U.S. territories in the Pacific Rim in the early part of the 20th century.

BELOW

Not until 1992 did the United States abandon its last military base in the Philippines. The base was located in Subic Bay, near Manila, in the upper left of this map.

★ POPHAM ★

Before the Pilgrims or the Puritans, There Were the Pophams.

In 1608, twelve years before the Pilgrims landed at Plymouth Rock, a man living in what is now Maine believed he'd be better off in old England. After building an oceangoing ship from scratch, he sailed to the British Isles to seek his fortune. His story may sound bizarre, but the facts are indisputable.

It all began in 1606, when King James of England granted the right to start colonies on the American east coast. The first group of colonists ended up near Florida, where they were captured by the Spanish. Oops.

The second group veered north, and in 1607 they landed in what is now Maine. They built a fort and hunkered down for the winter. Things were going reasonably well until the colony's leader, George Popham, abruptly died.

So in early 1608, twenty-five-year-old Ralegh Gilbert took over the "Popham Colony." The nephew of Sir Walter Raleigh, Gilbert led the group in the completion of their signature project: the construction of a seaworthy ship, built from local timbers. It was a demonstration project of sorts, proving that the New World was a good place for ship building.

By now you're likely wondering why the Popham Colony isn't as famous as the Pilgrims' Plymouth Colony. Blame Gilbert's family back in England. A bunch of them died, which meant that Gilbert inherited a big castle, located in Devon.

So Gilbert decided to sail back to England to claim his newfound wealth. Lacking a clear leader, the Popham colonists followed, thus ending the story of New England's first colony.

Had they stayed, the group might have enjoyed a history much like the settlers of Plymouth Colony, which surely would have resulted in a strikingly different map of New England.

OPPOSITE

The Popham Colony wasn't the first European settlement in the New World, but it predated the Pilgrims' arrival by more than a decade.

BELOW

Popham colonists chose a prime location for their settlement. The site has been home to a variety of military forts over the years, from the revolutionary era through World War I. It's a popular recreation spot as well. Today, the remnants of the colony are located within Popham Beach State Park.

Fort Popham, Me., Boston Boat Passing Fort Popham.

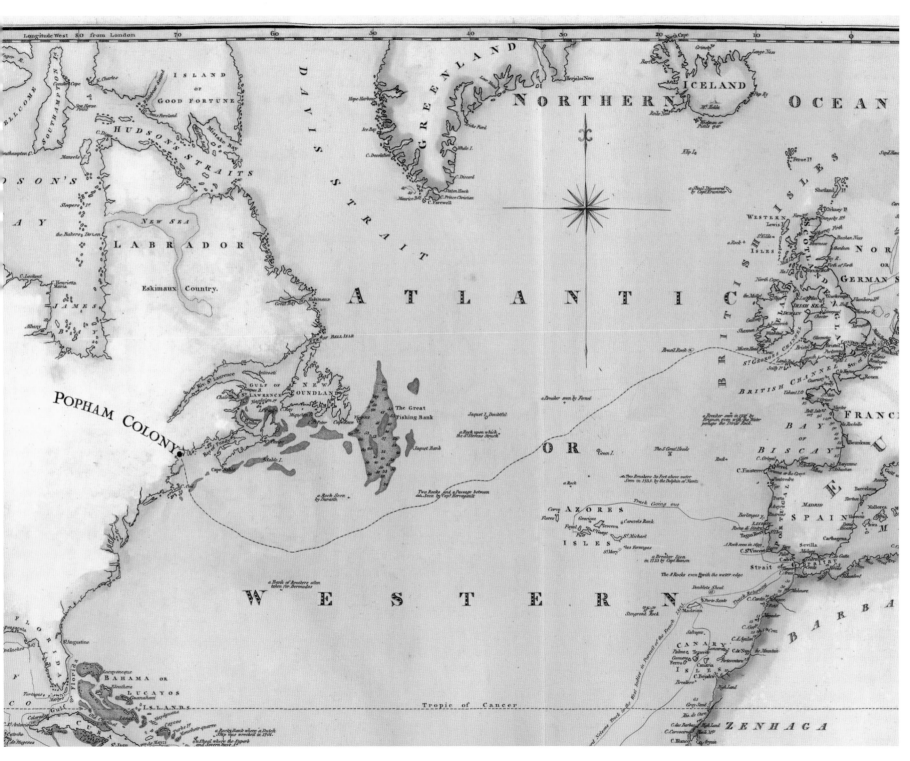

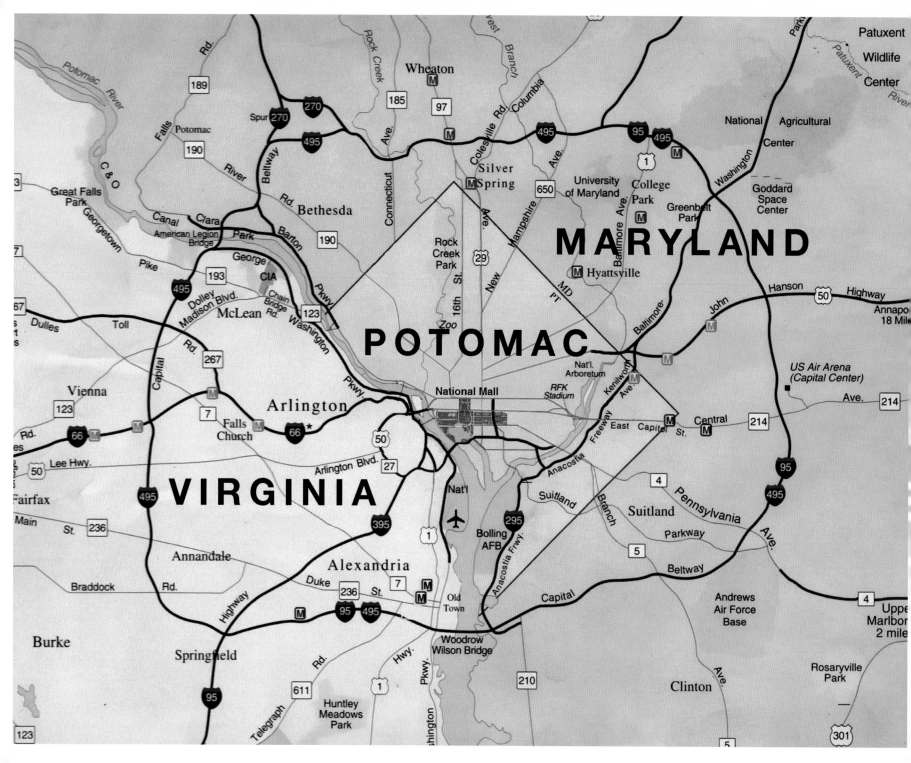

⋆ POTOMAC ⋆

Yes, it's Washington D.C.—but We Already Have a Washington.

The whole idea behind the creation of the District of Columbia was to ensure that the people running the federal government aren't hassled by any particular state. For example, if the District of Columbia didn't exist—and the U.S. capital was located in Maryland—imagine what would happen if Congress passed a law that Marylanders didn't like. They could retaliate by not picking up the garbage. Or if they really wanted to make Congress miserable, they could shut down all the bars and massage parlors.

Unfortunately, the founding fathers failed to give proper representation in Congress to citizens of the district. Residents have no senator or congressperson, and only recently have they been allowed to vote for president. Given that Americans are big on that whole democracy thing, this situation has made a lot of people unhappy over the years.

In 1982, residents voted in favor of creating the fifty-first state, but Congress did not approve. The district certainly has enough people to qualify for statehood; its population is about the same as those of Alaska, North Dakota, and Vermont.

The biggest obstacle is political. Since the district is heavily Democratic, adding it as a state would likely mean the addition of two more Democratic senators, a less-than-thrilling proposition for Republicans.

America experienced a similar situation in the late 1950s, when Hawaii and Alaska entered the union—essentially—as a pair. One was conservative, the other was liberal. Going forward, that's probably the only way new states will be added.

If Washington, D.C., ever nears statehood, Congress will have the curious problem of choosing a name. "Washington" is the obvious choice, but that's already taken. Lately, supporters have advocated "New Columbia," but I much prefer "Potomac." Many accounts claim that the name means "a place where people trade." Now that's perfect for Washington.

OPPOSITE

Proposed state of Potomac, with the famous "beltway" (I-495).

BELOW

Originally, the District of Columbia included land on both sides of the Potomac River. The south side was returned to Virginia in the mid-1800s.

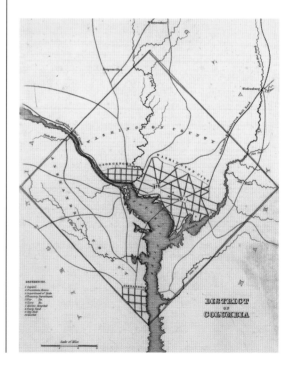

★ PUERTO RICO ★

Is a Beauty Pageant Standing in the Way of Statehood?

In my lifetime, Puerto Rico has flirted more closely with statehood than any other place. It certainly has enough people, boasting more citizens than twenty-four existing states. And the ratio of Puerto Ricans in favor of statehood has, at times, approached the 50 percent mark.

So what's the problem?

There was a time when language was cited as the major barrier. The argument was that Americans speak English; Puerto Ricans speak Spanish, and the two could not coexist. Now, of course, we think differently about such things. Puerto Ricans are learning English, and Spanish has become tightly integrated into life on the mainland. (I wasn't sure about this until I loaded up my Ford *Mustang* with a bowl of *guacamole* for a *Fiesta* Bowl party on my friend's *patio*.)

Today the bigger gulf is economics. Puerto Ricans may be better off than other residents of the Caribbean, but per capita income is much lower than in any of the fifty states. Statehood proponents suggest that poverty is precisely the reason that Puerto Rico should become a state; opponents cite it as a reason to exclude it.

Perhaps the most powerful proponent of all time was President Gerald Ford. Toward the end of his term, Ford decided to propose Puerto Rican statehood. What was weird was that no one expected it, not even the governor of Puerto Rico. Ford was probably just trying to pad his entry in the *World Book* encyclopedia. After all, his term as president wasn't exactly Rushmore-worthy.

For all we know, the core of the issue may have something to do with the Miss Universe pageant. If Puerto Rico becomes a state, it will no longer be able to enter its own contestant—and they've already won three times! Obviously, Puerto Rico must have an abundance of attractive and talented women. I suspect that is the real reason certain U.S. representatives have pressed for statehood.

OPPOSITE

Map showing Puerto Rico in relation to the U.S. mainland.

BELOW

The names on this map demonstrate the island's Spanish heritage. But heritage should not be a barrier to statehood. After all, the United States grabbed great big California from the Spanish, so why not teeny Puerto Rico?

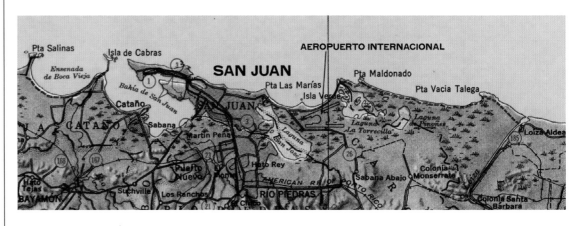

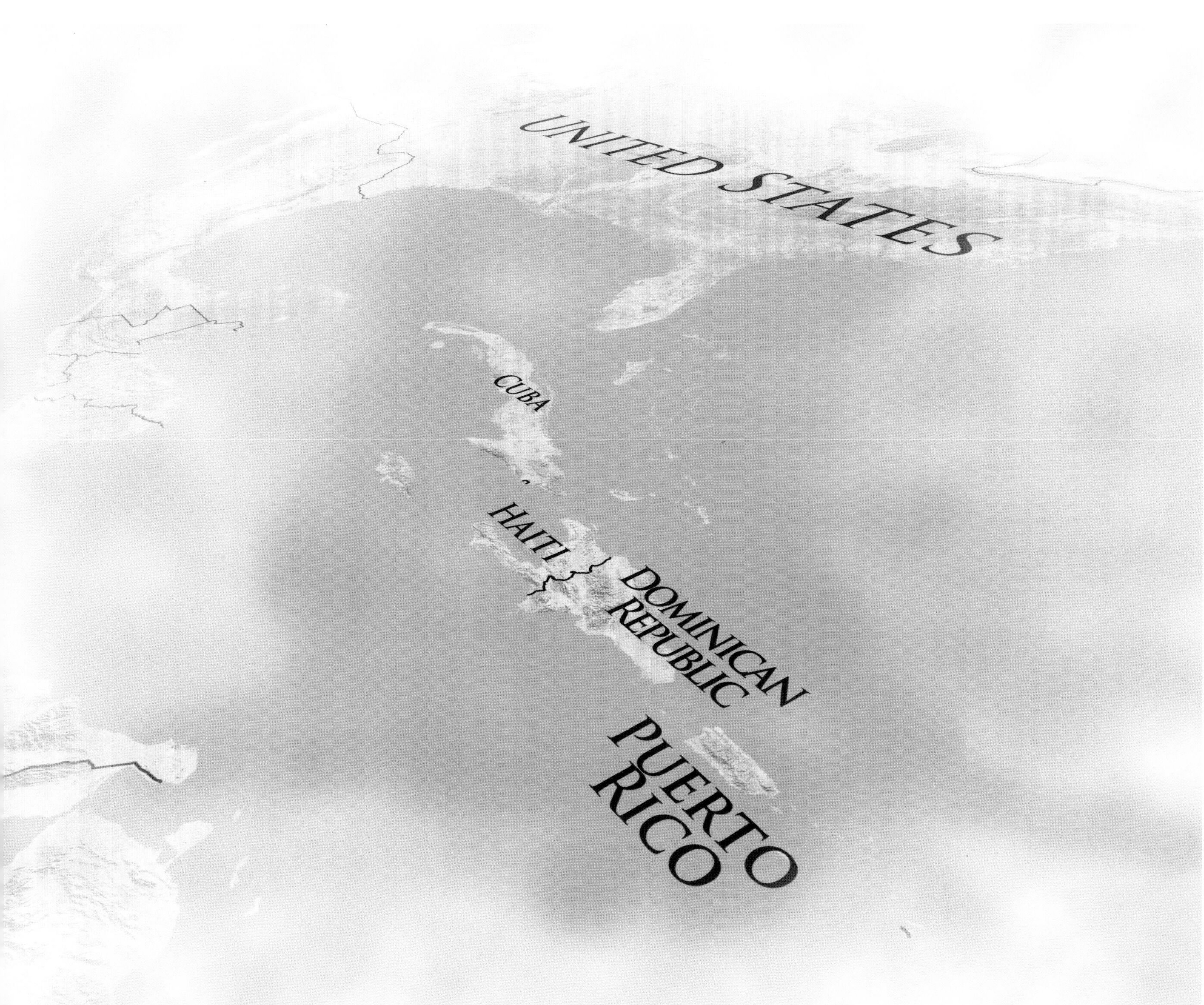
UNITED STATES
CUBA
HAITI
DOMINICAN REPUBLIC
PUERTO RICO

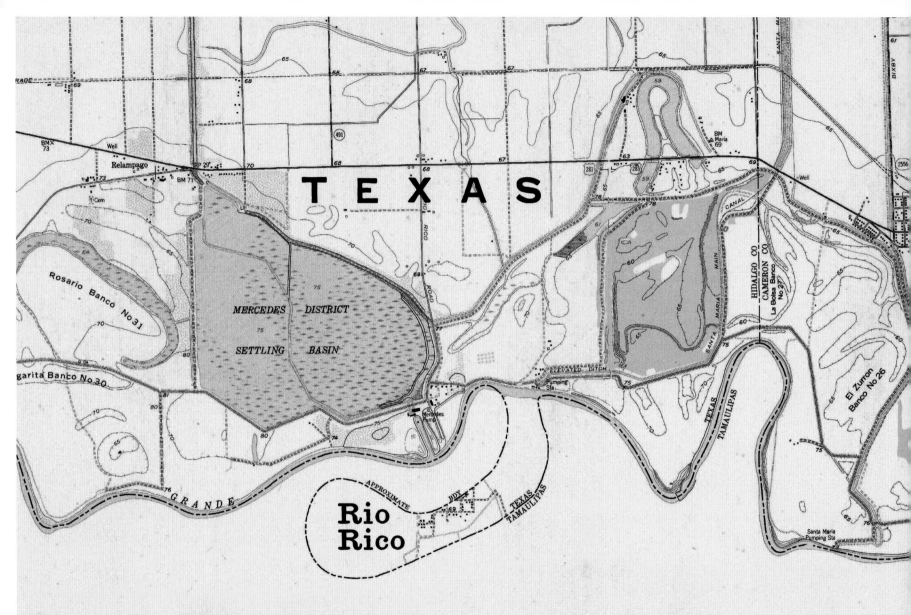

T E X A S

M E X I C O

Rosario Banco No 31

Margarita Banco No 30

MERCEDES   DISTRICT

SETTLING   BASIN

Relampago

Well

BMX 73

Cem

BM 71

RP 29

491

RIO RICO ROAD

Mercedes Pump

GRANDE

**Rio
Rico**

APPROXIMATE   BDY

TEXAS
TAMAULIPAS

ELEVATED   DITCH

Pumping Sta

TEXAS
TAMAULIPAS

SANTA MARIA MAIN CANAL

HIDALGO CO
CAMERON CO

La Bolsa Banco No 27

BM Maria

Well

281

BIXBY

2556

El Zurron Banco No 26

Santa Maria
Pumping Sta

★ RIO RICO ★

In Mexico or the United States? Depends When You Ask.

During Prohibition, Americans would do just about anything for a drink. And since alcoholic libations were still legal in Mexico, it's not surprising that Texans regularly waded across the Rio Grande, the river that serves as the boundary between the two countries.

This gave entrepreneurs A. Y. Baker and Joe Pate an idea. They would build a bridge near the Mexican town of Rio Rico to help thirsty Texans pass more easily across the border. More than eight thousand spectators turned out for the bridge dedication. Rio Rico developed into a major hotspot featuring dog racing, mariachi bands, even souvenir shops. It was rumored that mobster Al Capone had bankrolled the whole operation.

But this Mexican boomtown had one big problem: It was sitting on American soil. No one knew this at the time because the city was on the south side of the Rio Grande, and the river was the established dividing line between the United States and Mexico. What everyone had forgotten was that years earlier the river had been illegally diverted. Changing the river's course did not affect the official border.

The truth about the river's diversion wasn't rediscovered until 1967. Both the United States and Mexico quickly acknowledged that the border should be adjusted to put Rio Rico back in its rightful place, within the United States.

Doing so touched off a second boom, as pretty much everyone in Mexico claimed to have been born in Rio Rico, which would have made them instant American citizens. Federal law states that anyone born on American soil is automatically a U.S. citizen.

I should note that the Rio Rico tract is only about four hundred acres. So most of the thousands of birth claims were likely bogus, unless the gambling dens came equipped with birthing rooms upstairs. After sifting through all the claims, the Unites States did recognize more than nine hundred new citizens.

Too small to become its own state, Rio Rico was attached to Texas, making the second-biggest state just a tiny bit bigger.

OPPOSITE --

Modern U.S. geologic survey maps show a finger of America jumping the Rio Grande, to recapture lost territory.

BELOW --

In Brownsville, Texas—just a few miles from Rio Rico—officials display confiscated bottles of booze. Apparently, police procedure of the day called for pouring the illegal liquor on the street. Given the strength of prohibition hooch, I hope no one lit a match.

★ ROUGH AND READY ★

Like Animal House. Only with Gold Miners.

I know it seems preposterous that a few thousand gold miners could secede from California, but you have to realize that in the era of the California Gold Rush, pretty much anything was possible.

This all came into focus for me several years ago when I sat down with renowned Gold Rush historian J. S. Holiday. After the third beer, his thesis became clear: Young males tend to be good when people are watching, but when there's no authority around . . . well, things degenerate pretty quickly. In essence, the Gold Rush was like a nineteenth-century version of *Animal House*. I can almost hear the miners yelling, "Par-tee! Par-tee!"

You see, mining towns didn't have the usual authorities, like wives, pastors, or fathers-in-law. In fact, in the early months there wasn't even an official government. When governmental oversight did come to Rough and Ready in 1850, residents were outraged—for two reasons.

First, the county imposed a tax on mining claims, and nobody likes taxes. Second, and much more significant, was the prohibition of alcohol. Most Rough and Readians had originally come from Wisconsin, the state whose residents (both then and now) consume more booze than any other. One reason may be because Wisconsin was settled by beer-loving German Lutherans. (Beer is still so much a part of Wisconsin culture that it's not unusual for a Lutheran choir to tap a kegger after rehearsal.) That's why it's not surprising that the displaced Wisconsinites of Rough and Ready objected to their county going dry.

On April 7, 1850, residents declared independence, but rather than form a new state, they claimed to have created an entirely new country. That presented a problem no one had anticipated. Because they were no longer part of the United States, the citizens of Rough and Ready had no reason to celebrate the Fourth of July. These guys certainly did not want to miss a good party, so on the morning of July 4, 1850, Rough and Readians voted to rejoin the United States, and the whole matter was settled.

OPPOSITE --

The short-lived nation-state of Rough and Ready, as shown on a map from the era.

RIGHT --

This Milwaukee, Wisconsin, view dates from a couple decades after the Rough and Ready seces-

sion, but it explains the crux of the issue. Because German immigrants liked their beer, Milwaukee had a major brewery on nearly every city block. The desire for beer didn't change when residents raced off to California in search of gold.

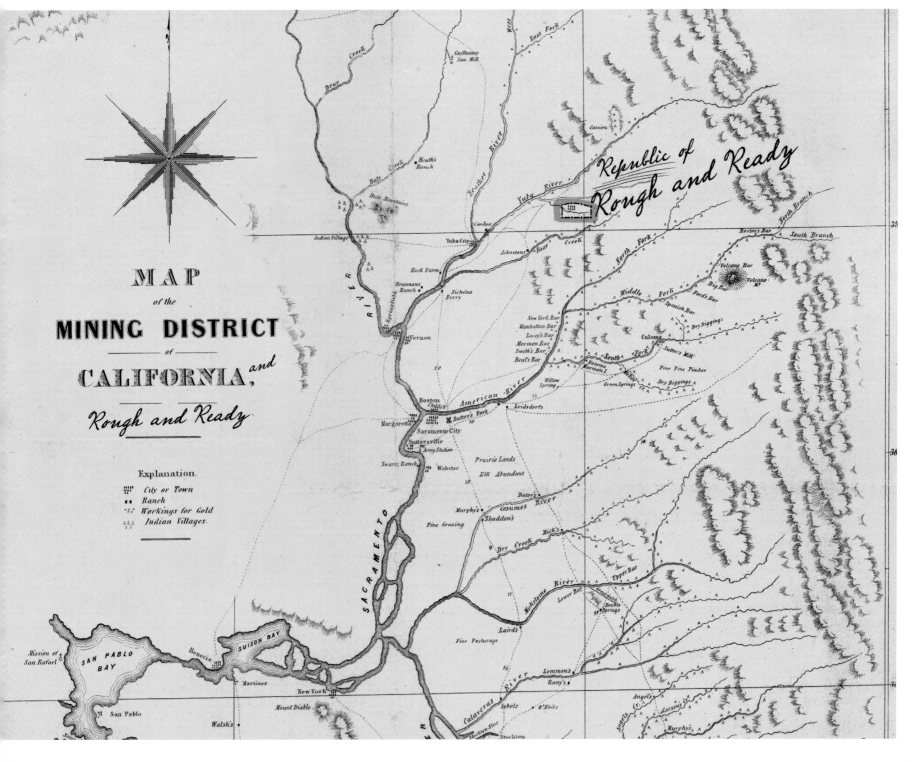

MAP
of the
MINING DISTRICT
of
CALIFORNIA, and
Rough and Ready

Explanation.
⊞ City or Town
∷ Ranch
⋏⋏⋏ Workings for Gold
▲▲▲ Indian Villages.

Republic of
Rough and Ready

Deer Creek
Creek
West
East Fork
Guillaume Saw Mill
Feather River
Dear
Heath's Ranch
Butte Creek
Camien
Yuba River
Butte Mountains
Cordua
North Branch
Rector's Bar
South Branch
Indian Village
Yuba City
Johnstons
Bear Creek
North Fork
Rock Farm
Brannans Ranch
Nicholas Perry
Middle Fork
Volcano Bar
Big Bar
Ford's Bar
Spanish Bar
Dry Diggings
Volcano M?
New York Bar
Manhattan Bar
Lacey's Bar
Merman Bar
Smith's Bar
Beal's Bar
Coloma
Sutter's Mill
Vernon
South Fork
Yatoma Mormon I.
Fine Pine Timber
Springfield
Willow Spring
Green Springs
Dry Diggings
RIVER
American River
Boston Childs F.
Leidsderts
Sutter's Fort
Margaretta
Saramento City
Suttersville
Army Station
Prairie Lands
Elk Abundant
Swartz Ranch
Webster
Murphy's
Batens
Cosumes River
Shadden's
Fine Grazing
Rich's
Dry Creek
SACRAMENTO
Mokelome River
Upper Bar
Lower Bar
Green Sprl.
Double Springs
Laird's
Fine Pasturage
Lemmon's
Rany's
Mission of San Rafael
SAN PABLO BAY
Benecia
SUISON BAY
San Pablo
Martinez
New York
Mount Diablo
Walsh's
Colaveras River
Isbels
O'Neils
Angel's
Stockton Slue
Angels Cr.
Carsons
Murphy's
Stockton

★ SAIPAN ★

Outpost of Shame. And Ralph Lauren.

Saipan is probably the most deplorable place in America. So why would a Bush administration official suggest that it should become the fifty-first state?

A bit of explanation: Saipan, an American island in the western Pacific, is part of the Northern Marianas group, a chain of islands that have been under American rule since 1898. Although the weather there may be beautiful, Saipan has been a nasty place to live in recent years due to a host of human rights abuses and a monster legal loophole.

Here's the skinny: Saipan garment companies import laborers from countries such as China and Thailand. The workers are subjected to horrible sweatshop conditions that the U.S. Department of Labor once called "slavelike." It's all legal, however, because Saipan has its own set of rules, stuff you could never get away with anywhere else in the United States.

If you own anything from The Gap, Liz Claiborne, Ralph Lauren, or Abercrombie & Fitch, it's possible your clothes were sewn in the sweatshops of Saipan. Checking the labels won't do any good, because garments from Saipan are "Made in the USA."

Eventually, the truth began to leak, and the U.S. Congress tried to pass laws to end the abuses. Saipan factory owners reacted quickly. Their well-paid lobbyists on Capitol Hill blocked the reforms. Today, conditions in Saipan remain pretty much unchanged. Wages are low. The local government is still corrupt. Human trafficking runs rampant. And so it goes.

Fast-forward to 2008, when David Cohen, the Bush administration's senior official in the region, expressed a desire to merge the Northern Marianas islands (including Saipan) with nearby Guam to create the fifty-first state. Admittedly Cohen made this statement after leaving office, but the governor of Guam immediately echoed the idea.

Lest we paint with a broad brush, Guam is nothing like Saipan. It's a solid democracy with a vibrant economy. They even elect a (nonvoting) representative to Congress. But the population is a little on the low side to justify statehood, hence the push to combine Guam with Saipan and the rest of the Northern Marianas.

Even the most optimistic statehood proponent agrees that the process would take a long time. For now, Saipan remains the American outpost of shame.

OPPOSITE

It's unlikely that an English-language poster would have been used to recruit workers in China or Thailand, but this image does summarize the pitch. Saipan was presented as a place with better wages and better conditions. It was all a lie.

BELOW

Aerial photo of Saipan. The sign may not be real, but the sentiment is.

★ SEQUOYAH ★

The State. The Inventor. Not the Tree.

In the late 1800s, many Native American tribes were forced from their homelands and relocated to the Oklahoma Territory. That's the broad-brush version of the story; the truth is more nuanced.

Starting in 1890, the land that's now Oklahoma was divided into two territories. At that time, only the western half was called "Oklahoma"; the eastern half was called "Indian Territory." As you might guess by the name, Indian Territory is where the U.S. government had shoehorned large numbers of Native Americans. Around the turn of the twentieth century, the tribes figured it was time to ask Congress to turn their territory into a state.

As a first step, leaders of the Cherokee, Creek, Choctaw, Chickasaw, Seminole, and Osage tribes held a convention. They drafted a constitution, elected delegates, and even debated suffrage for women. Statehood was put to the people for a vote and more than 80 percent of the 65,000 voters were in favor of the plan.

But the U.S. Congress could not be bothered. They never even considered the proposal. This really shouldn't come as a big surprise, given the federal government's long history of dissing Native Americans.

The proposed name, "Sequoyah," is itself an interesting story. Half-Cherokee (and perhaps half-white) Sequoyah wasn't a famous chief or renowned warrior. Rather, he was a quiet, disabled man who spent the better part of a decade working on a most remarkable invention: an alphabet. (Actually, it was a syllabary, but the basic idea is the same.)

Early on, Sequoyah realized the power of the white man's "talking leaves" (or writing). But tribal languages are oral, and so there was no way to write things down. Sequoyah set upon the task of creating symbols for each sound in the Cherokee language, and then he painstakingly built the entire language on paper.

Once the power of Sequoyah's invention was demonstrated, it quickly spread. Schools were opened just to teach this new form of communication. Today, nearly two centuries later, it's still in use by many Cherokee-language speakers in Oklahoma.

OPPOSITE

Sequoyah, superimposed on a map from the period.

BELOW

Sequoyah's Cherokee syllabary is still in use. Its distinctive font is preinstalled on many personal computers.

Cherokee Alphabet.

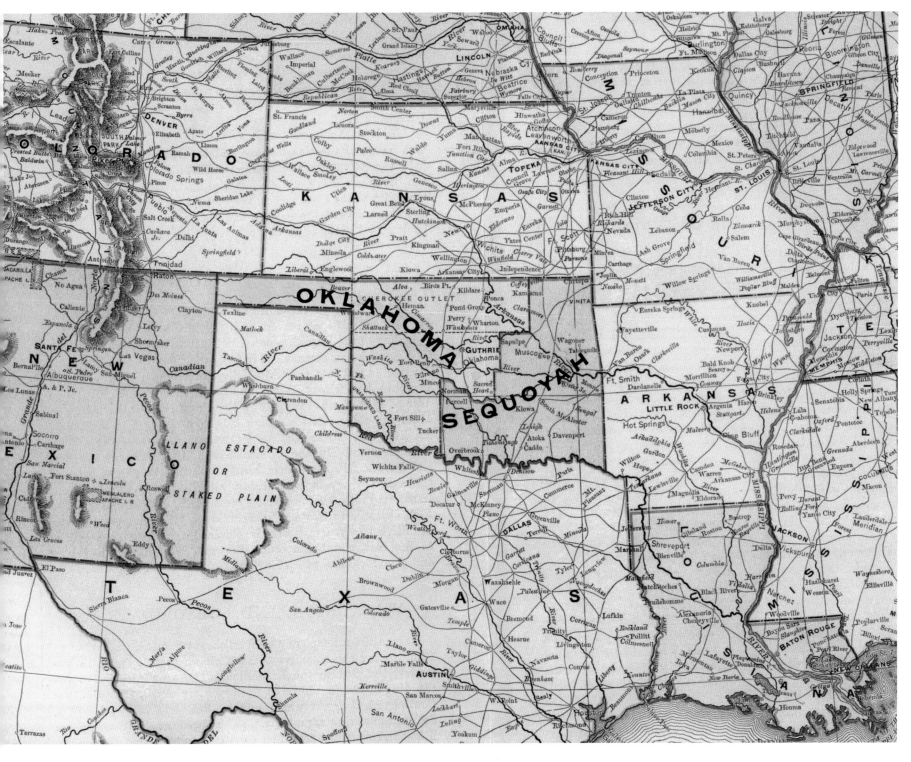

★ SHASTA ★

The Water Rush. Bigger Than the Gold Rush.

"This one has great water rights." That's what my real estate agent said as we tromped around a five-acre desert ranchette in eastern Idaho. I eventually bought the place, but I had no idea what water rights were.

I quickly learned.

In the American West, water rights are more valuable than gold. That's because most of the West is a desert, so if you want to grow something, you need to pipe in water from somewhere else. If you pour this piped-in water on your land, your crops will grow and you can make a ton of money. But the land that doesn't have access to the water pipeline is essentially worthless.

Given the value of water in California, residents in the northern part of the state, who enjoy a good supply, became increasingly concerned about their downstate neighbors, who seemed insatiably thirsty. In 1957, many in the water-rich north figured the only way to protect their resource was to form their own state, which they named "Shasta."

Shasta's secessionist leaders spouted some great rhetoric, claiming that their goal was "self-protection against southern California imperialists" and that they were "the victims of metropolitan colonialism." It sounds over the top, but the claims did hold a certain truth. Many of Southern California's leaders did (and still do) lust for the fresh water of the north. And if you doubt Los Angeles' audacious approach to water acquisition, consider a plan they drew up in the 1960s: to pump in water from Canada (see caption below). I swear they'd drain the Great Lakes if they could.

As you might expect, the more-populous south rejected the Shasta statehood proposal, and the idea soon died. But the forces that set it in motion remain in place. Nothing is more critical to Southern California than piping in more water.

OPPOSITE --

Shasta, as proposed.

RIGHT --

If you think Shasta-ites were paranoid about Southern California stealing all their water, consider the colossal plan proposed just eight years later by an organization called the North American Water and Power Alliance (NAWAPA). The project, which would have been the largest engineering feat in human history, involved collecting the water of Canada's northern rivers in a massive state-sized lake. The water would then have been pumped thousands of miles to the cities of California and the American Southwest. It sounds like a diabolical plot cooked up by some evil genius, but the plan had many influential supporters in high places. The biggest objections came from Canadians, who recognized it as a giant theft of their natural resources. If Canada weren't a separate country, this stupefying plan might actually have reached fruition.

★ SICILY ★

Because Statehood Beats Fascism. And the Mafia.

For hundreds of years, the island of Sicily was independent of Italy. Then, in 1860, the two were united.

It did not go well. Sicily's economy collapsed, creating a vacuum filled by the mafia and, later, the fascists. Over an eighty-year period, more than one half million Sicilians fled because of increasingly poor living conditions. So when the American military liberated the island during World War II, the locals were ecstatic. (Remember, this was a time when America was celebrated as a land of opportunity and prosperity.)

In the postwar era, borders everywhere were redrawn. In Sicily, this realignment was seen by many as an opportunity to break away from Italy. But that required a strong partner. Enter the United States.

Antonio Di Stephano, leader of the Sicilian Party of Reconstruction, figured the best way to cement a relationship with the United States was to apply for statehood. Di Stephano was no crackpot. As the leader of a party forty thousand strong, he echoed the era's overwhelmingly pro-American fervor. His party even used the Statue of Liberty as their official logo.

The life of my Sicilian-born grandfather offers a bit of insight into this affection for America. His family immigrated to Chicago around 1900; he was orphaned a few years later. At age sixteen, he proudly enlisted in the U.S. navy (he fibbed about his age) and served in World War I. In his twenties, he received a letter from Sicily explaining that he was entitled to significant land holdings on the island. To claim the property, all my grandfather had to do was move back.

That he would not do. Better to be a poor American than rich Sicilian.

OPPOSITE -------

Looking south, Sicily in relation to Italy.

BELOW -------

Sicilians celebrating right after American liberation in 1943. Of course they're drinking wine. What did you expect, beer?

★ SONORA ★

How Fifty Guys Started a State. Almost.

In the mid-nineteenth century, Mexico was weak. So a Californian named William Walker figured he could grab some Mexican territory and create his own country . . . or at least a new state.

In 1853, Walker, along with just forty-eight men, captured Baja California (the long peninsula off Mexico's west coast). Apparently, all you needed back then was a few dozen mercenaries, and you could carve out your own country.

Walker might have stayed in power, but he made the mistake common to land grabbers: He got greedy. Specifically, he tried to expand his empire to include the Mexican state of Sonora.

This arrogance really annoyed the Mexican government, and they finally rallied their troops and chased Walker out of the country.

Back in the United States, where mounting a private army to take over another country is illegal, Walker was put on trial. (This point is good to keep in mind should you ever get any ideas about attempting a coup in, say, Liechtenstein.)

Despite his obvious guilt, Walker was quickly acquitted. His exploits were widely popular, for many Americans saw Mexican land as ripe for the picking. They'd hoped he could snatch them some bargain-priced real estate.

If Walker had not overreached, it's possible that his support in the United States might have translated into statehood for his so-called Republic of Sonora. Remember: Texas was an independent nation that had become a state. Why not Sonora?

It should be noted that Walker's exploits were not unique. In the mid-1800s, lots of other mercenaries attempted similar takeovers in Central America. In fact, there was a name for this sport: *filibustering*. Today, of course, we know filibustering as a legislative tactic used in the Senate, but the term was originally defined as a sort of nation-grabbing.

And whatever happened to William Walker? After his acquittal, he tried another filibuster, this time in Nicaragua. He was more successful on that adventure and served as president there for a year, until the U.S. sent him home.

In 1860, Walker made one more attempt, this one in Honduras. By now the British were a major force in the region, and they wanted to be rid of the American troublemaker. They grabbed Walker, but rather than sending him home, the Royal Navy turned him over to the Hondurans, who summarily executed him. He was thirty-six years old.

OPPOSITE

Walker's Republic of Sonora, if it had become attached to the United States. Had he been content with just the Baja peninsula, he might have succeeded.

BELOW

William Walker: freebooter, mercenary—and dead at age 36.

★ SOUTH CALIFORNIA ★

Actually, They Wanted to Call It Colorado.

We've got a South Carolina and a South Dakota, so why not a South California? It's hard to imagine, but it really almost happened. The state legislature, the governor, the residents—everyone was on board. Votes were taken, and the matter was sent to Washington, D.C., for final approval.

But Congress wasn't listening. They were preoccupied with a tiny matter called the Civil War. It hadn't started yet—the year was 1859—but tensions were rising. Folks in Washington just weren't interested in California's plan to slice itself in half.

The 1859 proposal was authored by Andres Pico, a wealthy landowner. Pico was a Californio—that is, part of the Spanish culture that had first settled in the region in the 1700s.

Long before Americans arrived in the far West, Californio agriculture and trade were thriving. Their problem was gold, or rather the lure of gold. Once the first nugget was discovered in the region in 1848, hordes of settlers rushed to northern California. The Californios were quickly outnumbered, and that's what led Pico to push for separation.

Of course, Pico's plan wouldn't be the last attempt to split California; it seems there's a new plan every week. But Pico brought California the closest to separation; never again would a plan have such widespread support.

Pico's proposal suggested calling the new state "Colorado," which, by the way, ranks among the most-coveted state names ever. After southern California's rebranding attempt failed, the Arizona territory laid its plans to rename itself "Colorado." But the rectangular state to the north beat them to the prize.

OPPOSITE

The closest California ever came to division: Pico's 1859 plan.

BELOW

Los Angeles in the 1870s, looking east from Prospect Park. Note the Los Angeles River in the middle of the image. It's actually a river! Now, it's a cement trough.

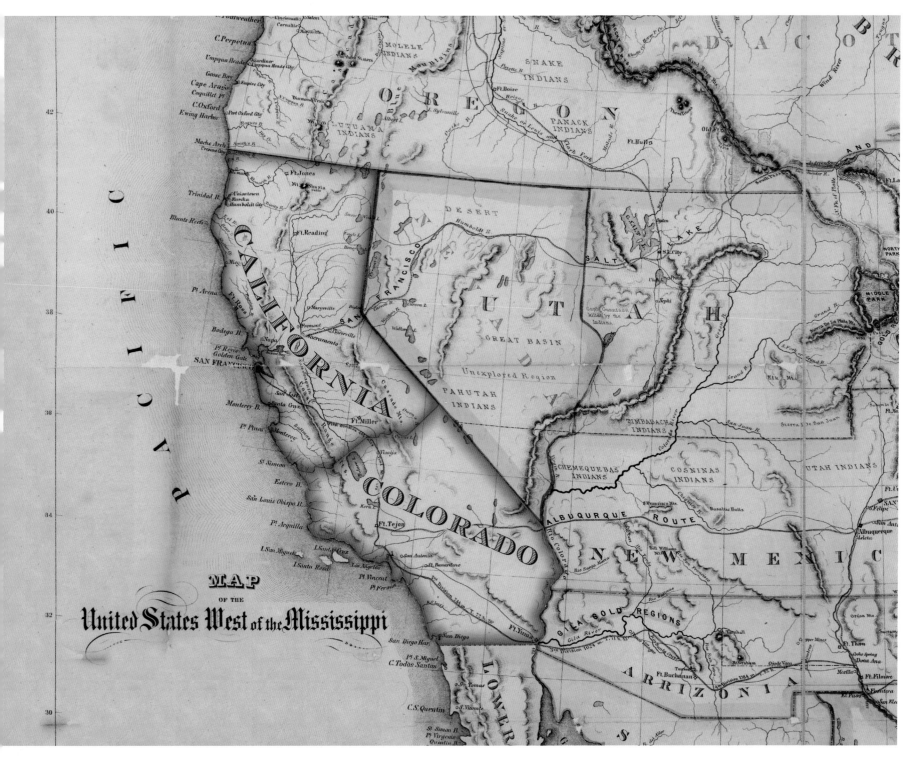

MAP
OF THE
United States West of the Mississippi

★ SOUTH FLORIDA ★

Mickey, Miami, and McMansions.

This story begins with something called "the big suck." It seems central Florida's Mickey-sprawl needed water—a lot of water. So a plan was created in 2008 to drain rivers in the northern part of the state and send the water southward.

You can imagine that this idea was pretty unpopular in the northern counties. Many there were already upset about tax increases, which were being used to rebuild south Florida's McMansions—you know, those monstrous estates that get blown over by the hurricane of the week.

Meanwhile, south Floridians had their own set of beefs. For one, they really wanted slot-machine gambling, but the conservative folks to the north refused to acquiesce.

By 2008, all this factionalism led a few south Florida legislators to threaten to form their own state. The proposal included the counties of Palm Beach, Broward, Miami-Dade, and Monroe.

When tempers eventually cooled, the proposal dissolved. But note that this wasn't the first time a piece of south Florida had threatened secession. In 1982, Key West seceded, renaming itself the "Conch Republic." Citizens then immediately applied for U.S. foreign aid.

They're still waiting.

As you may have guessed, the Conch Republic was nothing more than an artful publicity stunt, but you can buy realistic-looking passports for just $100 ($118 in Canada). According to legend, one of these passports saved the life of a Florida man after he was confronted by America-hating revolutionaries in Guatemala. He held up his Conch Republic passport and—instead of gunshots—received Tequila shots.

OPPOSITE

The borders of the state of South Florida, as proposed.

BELOW

Back when Florida's biggest tourist attractions were alligator wrestling and oddly posed women, Orlando's need for water was fairly modest. But in the years since Walt Disney discovered the region, the need for fresh water has become increasingly acute. Draining the northern counties remains an unpopular solution—at least in the north.

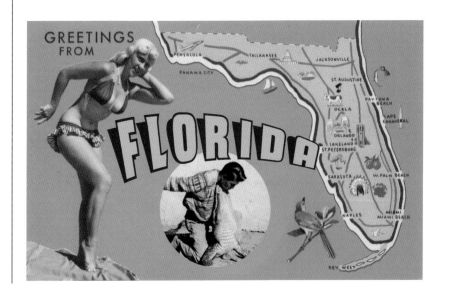

★ SOUTH JERSEY ★

An Old Idea That's Never Been Whacked.

Outsiders assume that New Jersey is all about grimy factories and chubby mobsters. The Garden State has those assets, of course, but there's more there than meets the eye.

Not long ago, I was driving through North Milford, New Jersey—just thirty-five miles from New York City—and was shocked to discover a landscape indistinguishable from Montana. I'm not exaggerating. New Jersey actually has a bear-hunting season. In fact, the state's Department of Fish and Wildlife has created a brochure outlining helpful tips for surviving a bear attack. (Top tip: bang pots and pans to make noise. Next time I'm strolling through the New Jersey wilderness with a backpack full of kitchenware, I'll give it a try.)

Admittedly, northeastern New Jersey is urban and asphalt covered. But the south and western parts of the state are surprisingly rural and unspoiled. It's this dichotomy that has led to three hundred years of attempts to divide the state in two.

As long ago as the late 1600s, New Jersey was split into eastern and western jurisdictions. The west side was home to orderly Quakers; whereas the eastern portion was populated by a varied mix of more cosmopolitan folks, such as the Puritans.

No one was ever quite sure where to draw the line between east and west, which is why the old map on the facing page shows multiple options. Even after New Jersey was united in the early 1700s, urban–rural differences never disappeared.

The push to split the state in two flared once again in the 1970s and '80s. The matter was put to a referendum, and voters in five of six southern counties approved a secession proposal. But then Thomas H. Kean was elected governor, and he began to redress some of the bigger issues inflaming South Jerseyans. The idea has since petered out—at least for the time being.

OPPOSITE ------------------------------------

(*Left*) Approximate line of the 1980 proposal to split off southern New Jersey to create a new state. The angled lines through the center of the archival map (*right*) show a few of the early proposals for splitting New Jersey in two.

ABOVE --

Yes, this is New Jersey. It's a land of stunning scenery, bucolic farms, and rural landscapes.

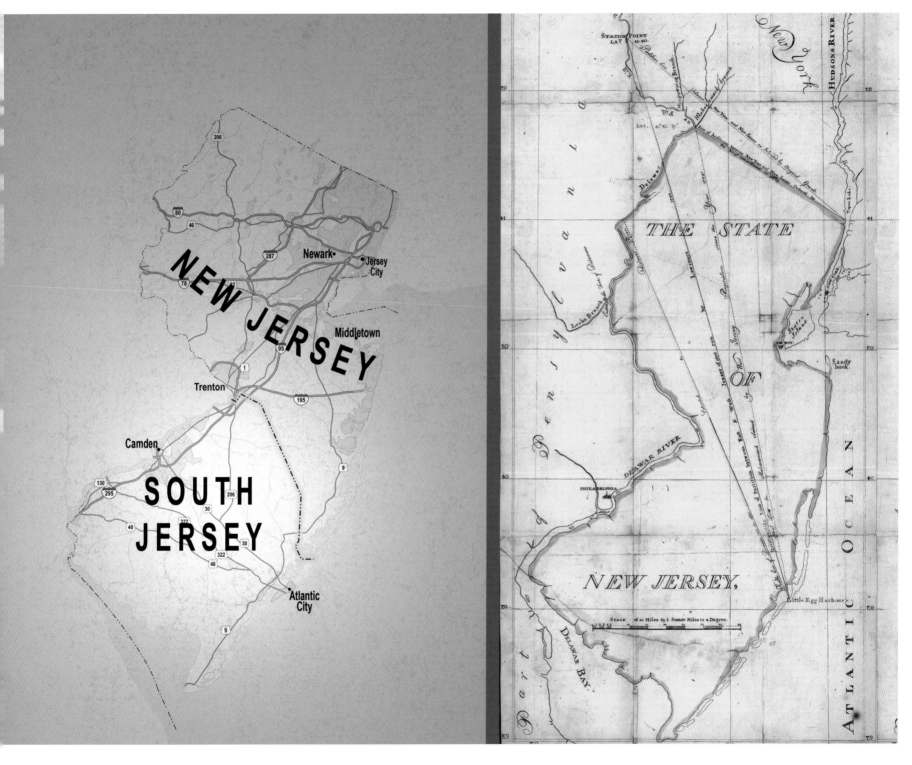

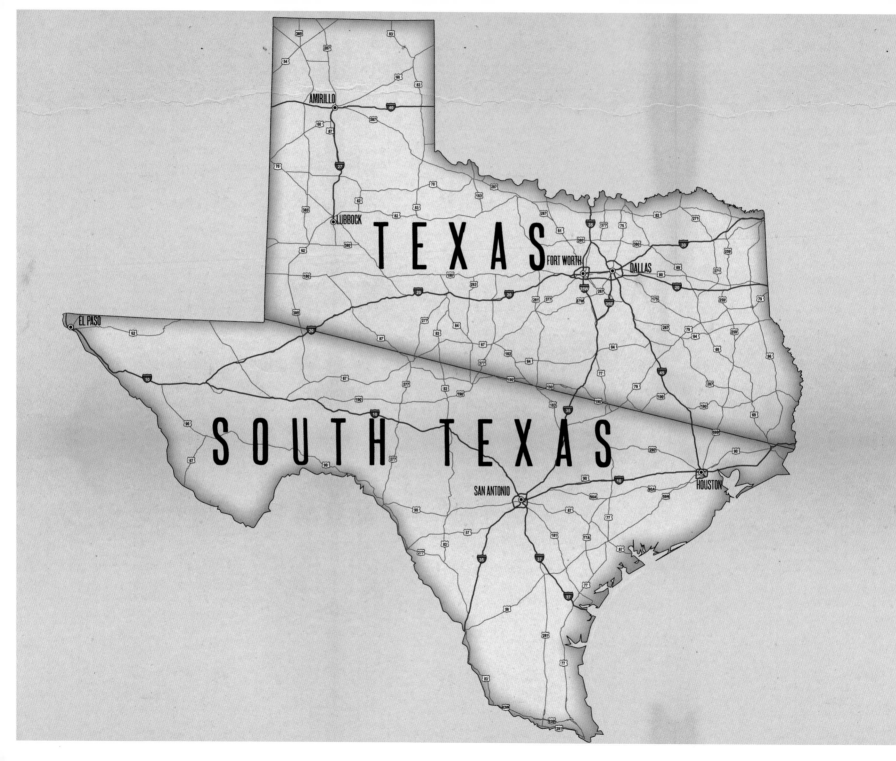

★ SOUTH TEXAS ★

How Many Texases Should There Be?

If you were paying attention in history class, you know that Texas was once a part of Mexico and later an independent nation. Then, in 1845, it joined the United States. But as a condition of statehood, Texas reserved the right to split itself into a whole bunch of states if it chooses to do so. Even today, the Lone Star State could slice itself up like a pizza.

Why would Texas even consider such a proposition? One advantage is that Texans would suddenly have four, six, or even ten senators in Washington, instead of just two. But the slice-and-dice idea has never gotten much traction, except for one episode in 1969, when Senator V. E. "Red" Berry proposed splitting the state north–south. His justification was that the north was mostly Protestant, white, and rich, whereas southern Texas was more Catholic, Hispanic, and poor. Nowadays, of course, the idea of separating groups of people for these reasons would be considered pretty offensive. But the constitutional amendments committee of the Texas state senate took the bill quite seriously, approving it for floor debate.

If the votes had gone in favor of Berry's bill, the deal would have been done: South Texas would be state number fifty-one, since the state doesn't require congressional approval to split up. But Berry's bill didn't get the necessary votes, and Texas remained whole.

Plans to split Texas have come up several other times. In 1930, Texas congressman John Nance Garner proposed the full monty—a five-state split. His goal: "To transfer the balance of political power from New England to the South and secure for the Southern States . . . prestige and recognition."

Wow, did he really think this was the best way to one-up New England? Wouldn't a Dallas Cowboys victory over the New England Patriots accomplish pretty much the same thing? At least then, Americans wouldn't have to go out and buy all new flags.

OPPOSITE

South Texas, according to Senator Berry's bill. San Antonio would be the capital of the new state, which would also include Houston.

BELOW

Some people think that Texas's ability to split is just some urban myth, but it's right there in the statutes of the twenty-eighth Congress, Session II, Res. 9, 10, p. 798.

ment of the United States. **Third.** New States, of convenient size, not exceeding four in number, in addition to said State of Texas, and having sufficient population, may hereafter, by the consent of said State, be formed out of the territory thereof, which shall be entitled to admission under the provisions of the federal constitution. And such States

★ STATE X ★

And State Y. A Lesson in Branding.

Reverend Charles Cummings wasn't your typical pastor—at least not by today's standards. There aren't too many pastors who bring a rifle to the pulpit. Ostensibly, Cummings did so because he was concerned about surprise attacks by the Cherokees, but maybe he just wanted to make sure no one fell asleep during his sermons.

In 1773, Cummings made his home in Washington County, Virginia, and soon took up a rather unusual hobby: statemaking. By 1784 he and a friend, Colonel Arthur Campbell, had rallied local support and petitioned Congress to create a new state that corresponds roughly to today's eastern Tennessee. Congress wasn't interested.

But the fine folks of Washington County weren't about to give up. Backed with additional support, they went back to Congress in 1785. This time, they had a two-state plan. The first overlaid most of modern Kentucky; the second covered what's now eastern Tennessee and northern Alabama (*see map, opposite*).

What's odd about this story is that neither Cummings nor Campbell could come up with a name for their proposed new states. Hence my "State X" and "State Y" designation. They really should have known that if you're trying to rally support, branding is everything. To sell anything, you've got to have an evocative name. For example, everyone wants to live in the "sun belt," but what if it were dubbed the "sweat belt." I think condo sales might drop a bit.

But Campbell and Cummings couldn't come with anything catchy. And you can't blame the era, because plenty of other folks cooking up new states had figured out the value of a good name (for example, Sylvania, Washington, Vandalia, and the like). In fact, two hundred years before ad agencies realized that adding "new" can make anything more appealing, colony-namers were already implementing the strategy. ("New Jersey. It's better than that old Jersey.")

So, despite filling out all the right paperwork—and outlining two quite reasonable states—Congress again rejected the proposal.

But Cummings and Campbell weren't done yet. The third time around, they tried to hook up with a nearby group that *had* chosen a memorable name: Franklin. The idea got a bit closer to passing muster with Congress, but in the end it fell short yet again.

I did find birth records of Reverend Cummings's children, which provide further evidence of his complete inability to come up with creative names. He named his son John and his daughter Mary.

OPPOSITE --------------------------------

States X and Y, superimposed on boundaries of a modern map.

BELOW --------------------------------

Rev. Cummings could have used this book. Too bad it didn't exist.

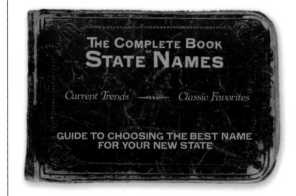

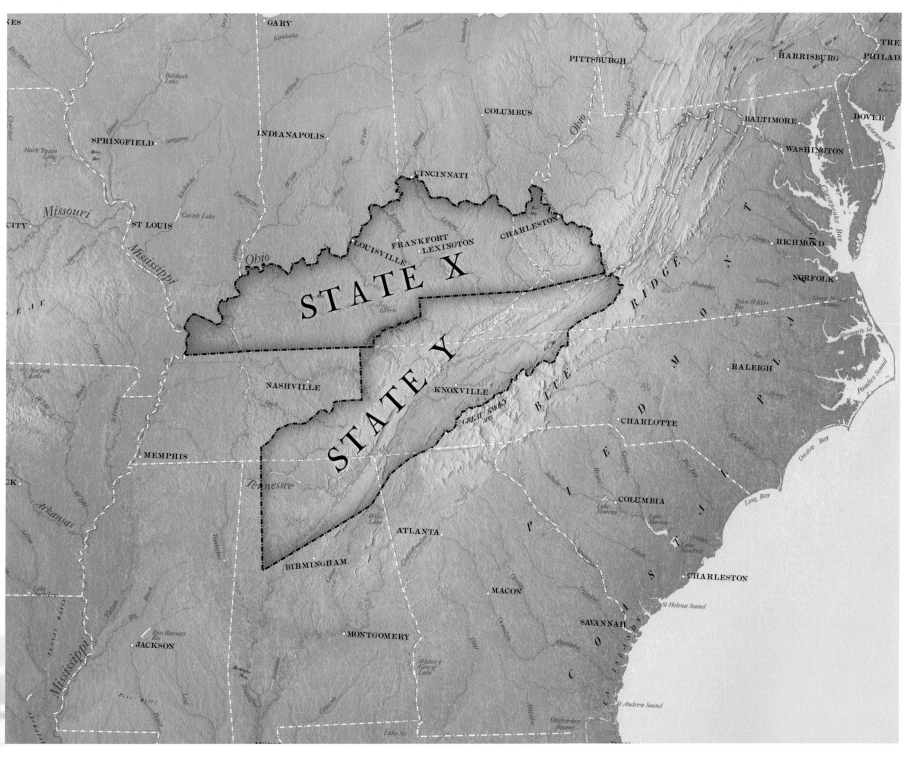

★ SUPERIOR ★

If You Treat Someone as Inferior, They Might Try to Become Superior.

Back in the 1830s, Michigan didn't want the Upper Peninsula—that's how this story begins. What Michigan did want was a bit more of the land that now forms Ohio. In the end, Michigan lost the disputed "Toledo Strip," so Congress figured they would console Michiganders by giving them the Upper Peninsula. It wasn't much of a consolation prize. The cold, inaccessible tract seemed worthless—it wasn't even connected to the rest of the state.

As you might imagine, people living in the Upper Peninsula felt a bit alienated. Heck, they couldn't even get to Lower Michigan much of the time. It wasn't until the 1950s that a bridge finally linked the two halves.

Even with modern bridges and roads, the distances from one end of Michigan to the other are still immense. For example, the city of Ironwood in the Upper Peninsula is six hundred miles from Detroit. Washington, D.C., is closer to Detroit than Ironwood is. Or, to put it another way, if Ironwood residents need to do a little shopping, Des Moines, Iowa, is closer than Detroit. So is Sioux Falls, South Dakota.

Given the geography, you might think it would make sense for the Upper Peninsula to hook up with Wisconsin, but the truth is that most of Wisconsin doesn't have much in common with the area either. Lower Wisconsin is a land of big blocks of cheese, endless corn fields, and some pretty big cities. The Upper Peninsula doesn't have any of that. Instead, it has rocks. Some of those rocks are quite beautiful, and some are very valuable—chock-full of copper and iron.

Over the years, there have been several attempts to split off and form a new state, which is usually dubbed "Superior." Talk was pretty serious as recently as the 1970s, when *Newsweek* magazine even ran an article on the idea.

Many Superior statehood proposals include parts of northern Wisconsin. My map bolts on several northwoods Wisconsin counties that are compatible with the Upper Peninsula.

Like many fifty-first state efforts, the big barrier for Superior would likely be population, or, more specifically, a lack thereof. Nobody lives up there. Even if you add in northern Wisconsin, the total population likely falls under a half million. It's hard to justify two new senators for a state that has fewer people than Boise, Idaho.

However, there is one scenario that might play into Superior's hands. Since Congress likes to add states in pairs (one Democratic and one Republican, to avoid tilting the balance of power), Superior could be the Republican counterweight to Democratic-leaning Puerto Rico or Washington, D.C.

OPPOSITE

This Superior map includes all the Upper Peninsula plus several counties from northern Wisconsin.

BELOW

When rocky cliffs and water meet, you have a problem. Ships tend to crash, especially at night. Thus, the coast of Lake Superior is dotted with scenic lighthouses, like this one.

★ SYLVANIA ★

Plus Polypotamia, Pelisipi, and a Bunch of Others.

Every student of American history knows that the Northwest Ordinance was among the most important pieces of legislation ever signed. It established that new states (starting in the Great Lakes region) would have all the same rights as the original thirteen.

But what hasn't been widely acknowledged is Thomas Jefferson's brilliant proposal to slice up this new territory. In 1784, Jefferson pitched a detailed plan to carve new states out of the Great Lakes territory. He even created a map of his proposed states. Congress didn't bite. Three years later, however, Congress revisited the idea and this time passed the famed Northwest Ordinance. But the final legislation did not include the specifics of Jefferson's map.

And therein lies the tragedy. Because, in my opinion, Jefferson's states are actually better designed than the ones that exist today.

Consider Jefferson's Sylvania. It would include what is now northeast Minnesota, northern Wisconsin, and Michigan's Upper Peninsula. Anyone who has visited this region knows that these areas fit together nicely. In fact, recurring efforts have proposed the creation of a fifty-first state called Superior (see page 133), which would closely follow Jefferson's plan.

Another example is Jefferson's Michigania, which is better designed than the state that now occupies the same space—Wisconsin. Wisconsin's industrial south differs greatly from the north woods. About all the two areas share is a love of the Green Bay Packers. Jefferson drew his border at the perfect dividing line.

Jefferson's plan also solves the Chicago vs. downstate Illinois problem by creating a horizontal state in the region. (Of course, Chicago didn't even exist in Jefferson's day.)

And Jefferson seemed to have a prescient understanding of the future tensions in Ohio, conflicts that would have been minimized under his plan. As recently as 2007, for example, the *Cleveland Plain Dealer* was proposing a state called Clevelandania, which follows Jefferson's idea.

If Jefferson was guilty of a misstep, it was in his choice of names for the new states. What was he thinking with "Polypotamia"? And I don't think any state should start with "Ass—."

The names weren't all awful. "Illinois" actually made it. And "Michigania" is pretty close, although it ended up as label for the east side of the lake, not the west.

Remember, Jefferson never visited the upper Midwest. All he had to work from was a map and scattered reports. Almost magically, he drew the perfect lines.

OPPOSITE

Jefferson's map, with the names he dreamed up.

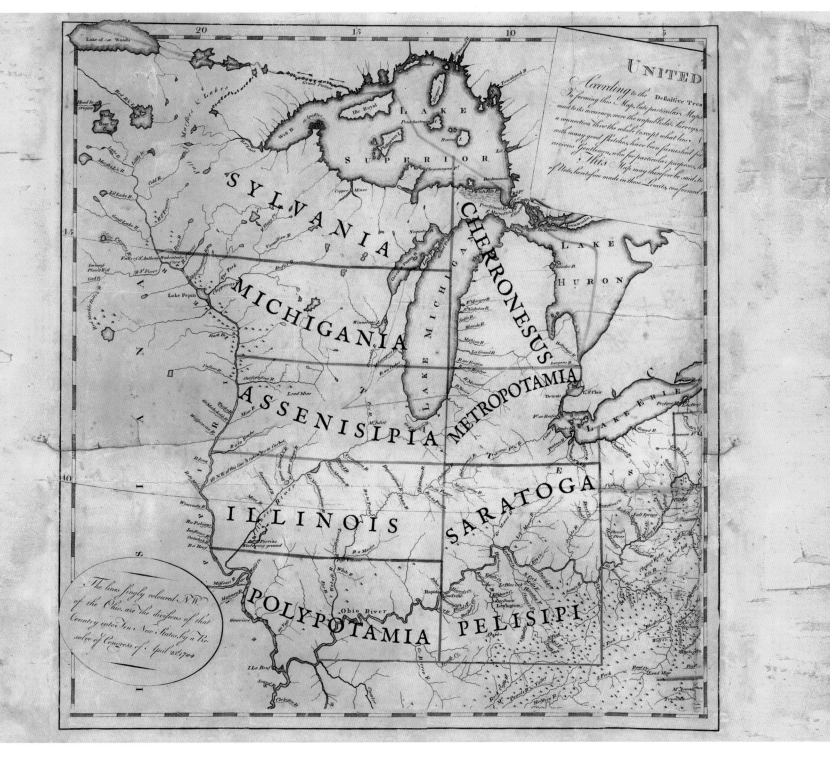

★ TEXLAHOMA ★

Another Candidate for the Forty-Ninth State.

In the early twentieth century, everyone was falling in love with the automobile. The new-fangled machines were fun, useful, and increasingly affordable. But cars need good roads, and in pre–World War II America, the roads were awful. People in rural places such as northern Texas and western Oklahoma were especially desperate for decent roads, but politicians in the faraway state capitals weren't listening.

And so the Texlahoma proposal was created. As designed by Oklahoman A. P. Sights, forty-six counties in Texas and twenty-three in Oklahoma would join to form Texlahoma. Legislators of the new state could focus on building better roadways—and all the other services that these "forgotten" counties believed they weren't getting.

The proposal did get some traction. Sights claimed that two-thirds of the politicians he polled favored the idea. The *New York Times*, although skeptical, covered the plan in significant detail. In addition, the American vice president at the time, John Nance Garner IV, was a big supporter of carving Texas into new states. A Texan himself, Garner probably just wanted more senate representation for his home state.

Then again, Garner did have a penchant for unusual causes. He zealously championed naming the prickly pear cactus as the Texas state flower. He lost that vote (to the bluebonnet) and was thereafter dubbed "Cactus Jack," which has to be the coolest nickname of any U.S. vice president.

The chief problem with all the proposals to split Texas was that the residents of the new state would no longer be Texans. That's a huge stumbling block to a population fiercely proud of its heritage. For that reason—as much as any other—Texlahoma never came to be.

OPPOSITE --

Texlahoma, as proposed.

TOP RIGHT --

Vice President John Nance Garner, a strong supporter of splitting Texas into new states.

BOTTOM RIGHT -------------------------------------

Garner's other quixotic cause was making the prickly pear cactus the state flower of Texas. It lost out to the bluebonnet—and Texlahoma fared no better.

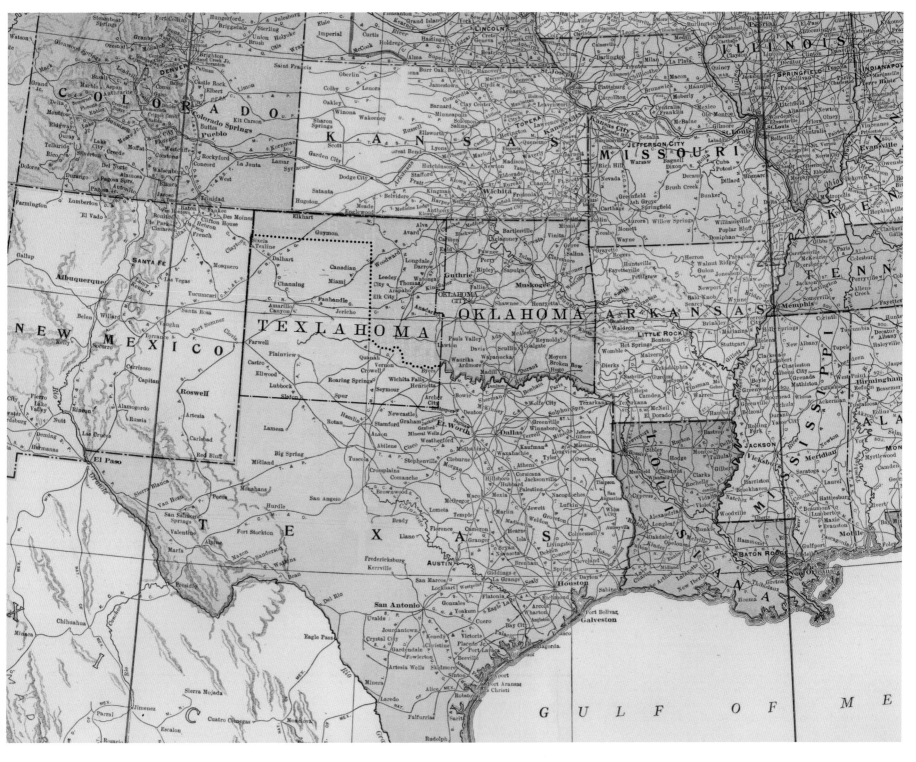

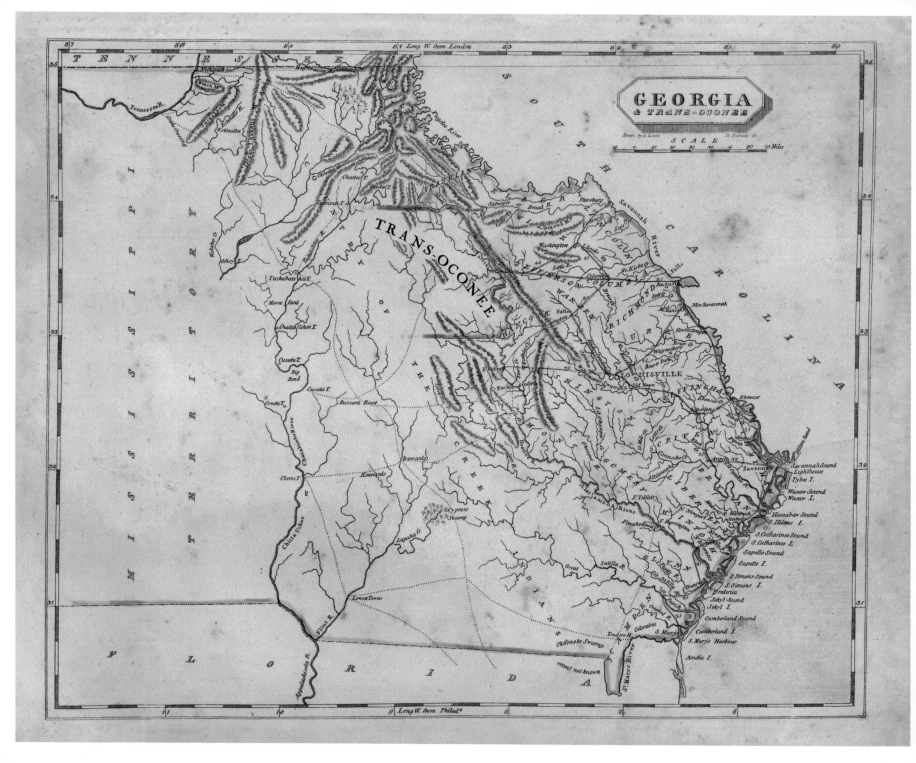

★ TRANS-OCONEE ★

A story of George, George, and Georgia. And Elijah.

Rule Number One when creating a freelance state: Don't tick off George Washington.

Elijah Clarke did not understand this rule.

Back in 1794, Clarke hatched a plan to settle the region west of the Oconee River in Georgia. By treaty, the land belonged to the Creek Indians. Clarke didn't care.

So in May of that year, Clarke and his private militia seized a chunk of Creek land and set up their own republic. It might have succeeded as a new nation, or more likely as a new state, but there was one big problem: President George Washington was dead set against it.

Washington saw Clarke's endeavor as clear violation of the Treaty of New York, signed by the United States and the Creek people. Of course, it's no secret that the United States often ignored such treaties, but Washington had signed this one, and he wasn't about to look the other way.

It may be hard to fathom today, but Washington was more than just a president. He was the most popular and powerful man in the world. To put it in modern-George terms, imagine the likeability of George Clooney coupled with the power of President George W. Bush. Then multiply by ten.

So when Washington leaned on Georgia governor George Mathews to bring an end to Clarke's enterprise, it became clear the nascent state was doomed. Mathews sent a militia of 1,200 against Clarke's outpost. Clarke vowed to fight to the death. But when Mathews offered amnesty, Clarke promptly surrendered, ending the whole affair.

BELOW --

Most portrayals of George Washington show him in a mythic, larger-than-life setting, such as this

one at Valley Forge. Still, there's no denying he was America's first rock star. When he wanted the Trans-Oconee movement squashed, all he had to do was say the word.

OPPOSITE --

This map approximates the borders of the Trans-Oconee Republic.

★ TRANSYLVANIA ★

No Vampires Were Involved. Probably.

Yes, Transylvania is the name of a place in eastern Europe where Dracula is said to have lived. But that's not what we're talking about.

The American Transylvania overlays Kentucky. And the name isn't as odd as it might seem. The word *sylvan* means "a pleasant woodsy area." It was a popular suffix during the colonial era; think of Pennsylvania, the pleasant woodsy region originally owned by William Penn.

Back in the early 1770s, most Americans lived east of the Appalachian range. The mountains were hard to cross, and, frankly, the western lands were kind of scary.

Then Daniel Boone enters the story. He offered Cherokee leaders a couple wagonloads of guns, clothes, and cooking utensils in exchange for a huge tract located west of the Appalachians. Just getting to this new land was a burden; in fact, several in Boone's company got scared and ran home, while others died in skirmishes with the native tribes. But in 1775 the expedition arrived on the Ohio River and established the town of Boonesborough.

The Transylvanian settlers set up a government of sorts and then sent a representative named Jim Hogg to the Continental Congress. But the Continental Congress wasn't too excited about adding any freelance states. Virginia especially was opposed, since it claimed the same land.

An interesting aside: One of Boone's first acts in Transylvania (now Kentucky) was to establish special rules for breeding horses, thus launching the beginning of the region's fame as equestrian country. Just think: If things had gone a bit differently, each spring horse aficionados might be enjoying the Transylvania Derby.

Anyway, when the Continental Congress failed to recognize Transylvania, the idea died. Until a few years later, when a different set of folks formed Kentucky.

OPPOSITE ------------------------------

The proposed state of Transylvania, superimposed on a map of the current states. It would have been the 14th state.

RIGHT ------------------------------

Charles Eugene Boone, better known as Pat. The famous singer was the great-great-great-great-grandson of Daniel Boone, the man who tried to establish Transylvania as the 14th state. (Some amateur genealogists cast doubt on Pat Boone's heritage, but he told me this story personally, and I believe it.)

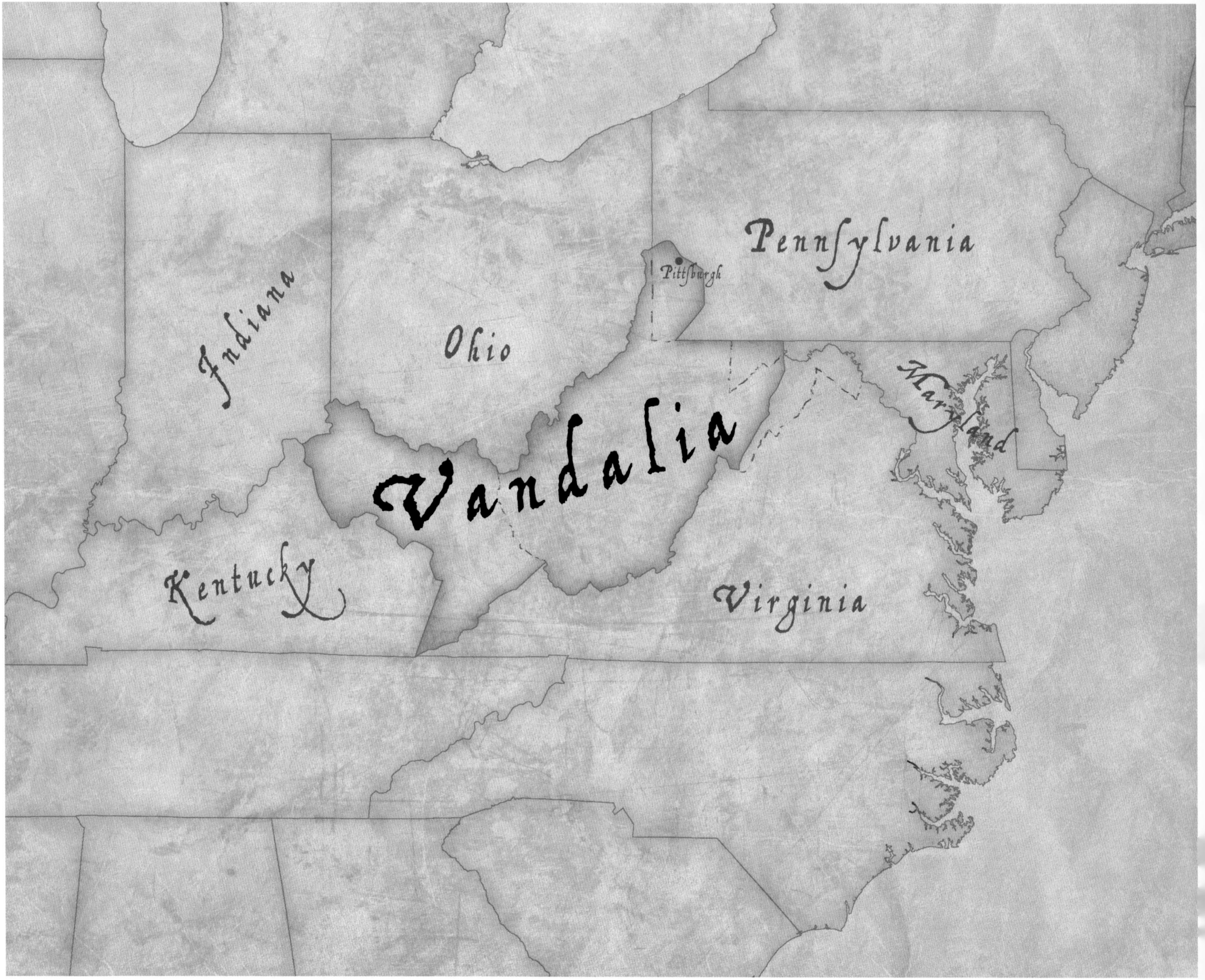
Indiana
Ohio
Pennsylvania
Pittsburgh
Vandalia
Maryland
Kentucky
Virginia

⋆ VANDALIA ⋆

When West Virginia Included Pittsburgh. Sorta.

Everyone knows that West Virginia split off from Virginia during the Civil War era. But what most people don't realize is that, by that time, the idea had already existed for more than one hundred years.

As early as the 1740s, opportunists were looking for ways to turn a profit from the Ohio Valley—and, more specifically, from the land around what is now Pittsburgh. But there was a problem: an Ottawa chief by the name of Pontiac.

For years, French settlers in the Great Lakes region had gotten along nicely with the local Native Americans, partly because they gave lots of free stuff to the tribes. Today we'd call it foreign aid, and the idea was basically the same then as it is now: Give a nation money, and its people will be nice to you. But when the British drove out the French, the Brits refused to continue supplying aid to the tribes. This angered Pontiac and his people, triggering a war.

Soon Native Americans in the area began a series of attacks on the British. In one of the battles, a group of traders lost a whole bunch of merchandise to Pontiac. One should probably expect this kind of thing in a war, but the merchants were bent on restitution. In a great PR move, they rebranded themselves as the "suffering traders" in 1763, and they just wouldn't shut up about how much they had lost (tens of millions of dollars' worth, in today's money).

Eventually, to compensate for their loss, the suffering traders were given a huge tract of land—Vandalia. A few years later, Vandalia's settlers asked the Continental Congress to recognize the area as a new province. By then, they had come up with a better name: Westslyvania.

Why the name change? Vandalia was a nod to the British queen, who was descended from Vandals. But if you wanted to impress the Continental Congress in 1776, naming your new state after the Queen of England was not the way to go. (Remember, the colonies were on the brink of the Revolutionary War.) So the settlers switched to the new and improved name: Westsylvania. None of this mattered, however, since Virginia and Pennsylvania claimed the same land as Westsylvania, and they blocked recognition.

About one hundred years later, the idea came up again. This time it succeeded, under the name West Virginia.

★ WASHINGTON ★

A State That Could Have Prevented the Iraq War.

In 1784, Congress drew the borders for a proposed state called "Washington." But this version of Washington was nowhere near the Pacific Northwest; rather, it encompassed much of what is now eastern Ohio.

What's interesting about the proposal is that it would have solved the biggest problem faced by Ohio today: an overabundance of big cities. Cleveland, Cincinnati, and Columbus are all top-forty metropolitan areas, but their conflicting needs can't be met by a single state government. The 1784 version of Washington would have averted the problem by separating Cleveland from the others. That makes sense, because Clevelanders tend to think and vote differently from their downstate neighbors.

If all this seems irrelevant, consider the following: If the proposed state had become a reality, Ohio's vote in the 2000 election would likely have been split. That means no George W. Bush presidency and probably no Iraq War. Seemingly arcane border issues can have a major impact on history.

Then there's the topic that's even more important to many people: professional sports. Columbus is among the biggest cities in America without a major sports team (unless you count hockey . . . or major-league bowling). That's because Columbus is in the same state as sports-rich Cleveland and Cincinnati. But stick a new state line in there, and a new sense of identity would be created, along with potential for fresh sports rivalries. Back in the early 1920s, Columbus had a really good NFL team, the Panhandles. Redraw the state line and the team would have a reason to return—a great new rival for the Browns and Bengals.

Admittedly, the NFL has no plans for expansion in Columbus, but the original Washington statehood idea does have its modern-day proponents. After a 220-year hiatus, the proposal was resurrected in 2007 by *The Cleveland Plain-Dealer* newspaper. Columnist/historian Thomas Suddes suggested a trimmed-down version, called "Clevelandia," that would carve a fifty-first state from northeast Ohio. Clevelandia would contain nearly four million people—plenty for a new state.

And the state song suggested by Suddes would have instantly become best in class: "Cleveland Rocks."

OPPOSITE

The 1784 plan for the state of Washington.

BELOW

Pro football's Columbus Panhandles are unlikely to make a comeback, at least not as long as Cleveland remains in Ohio.

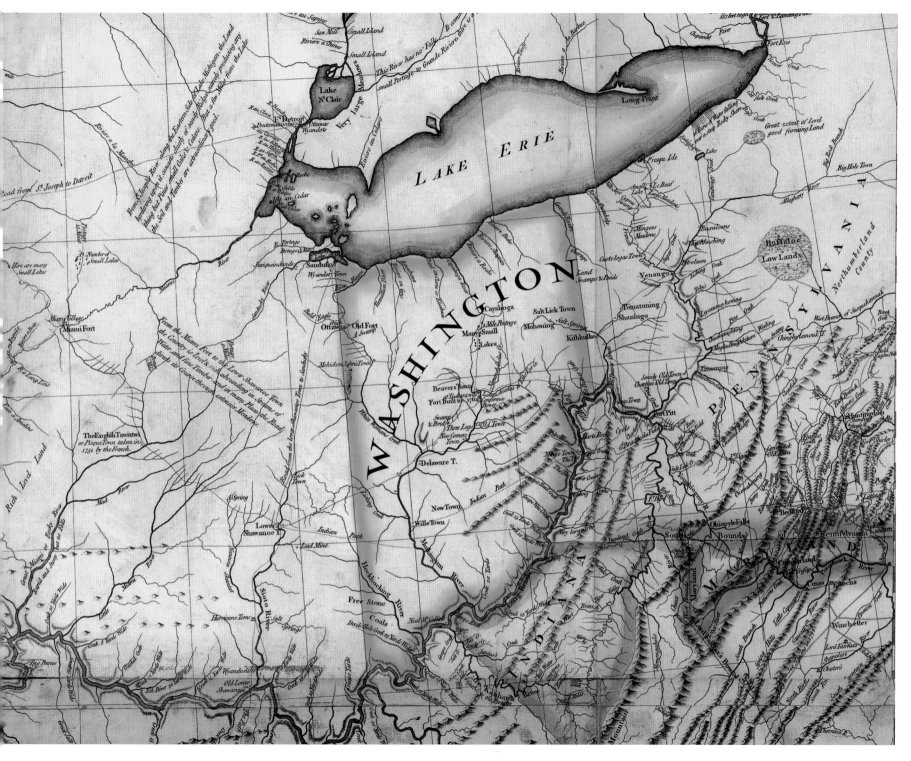

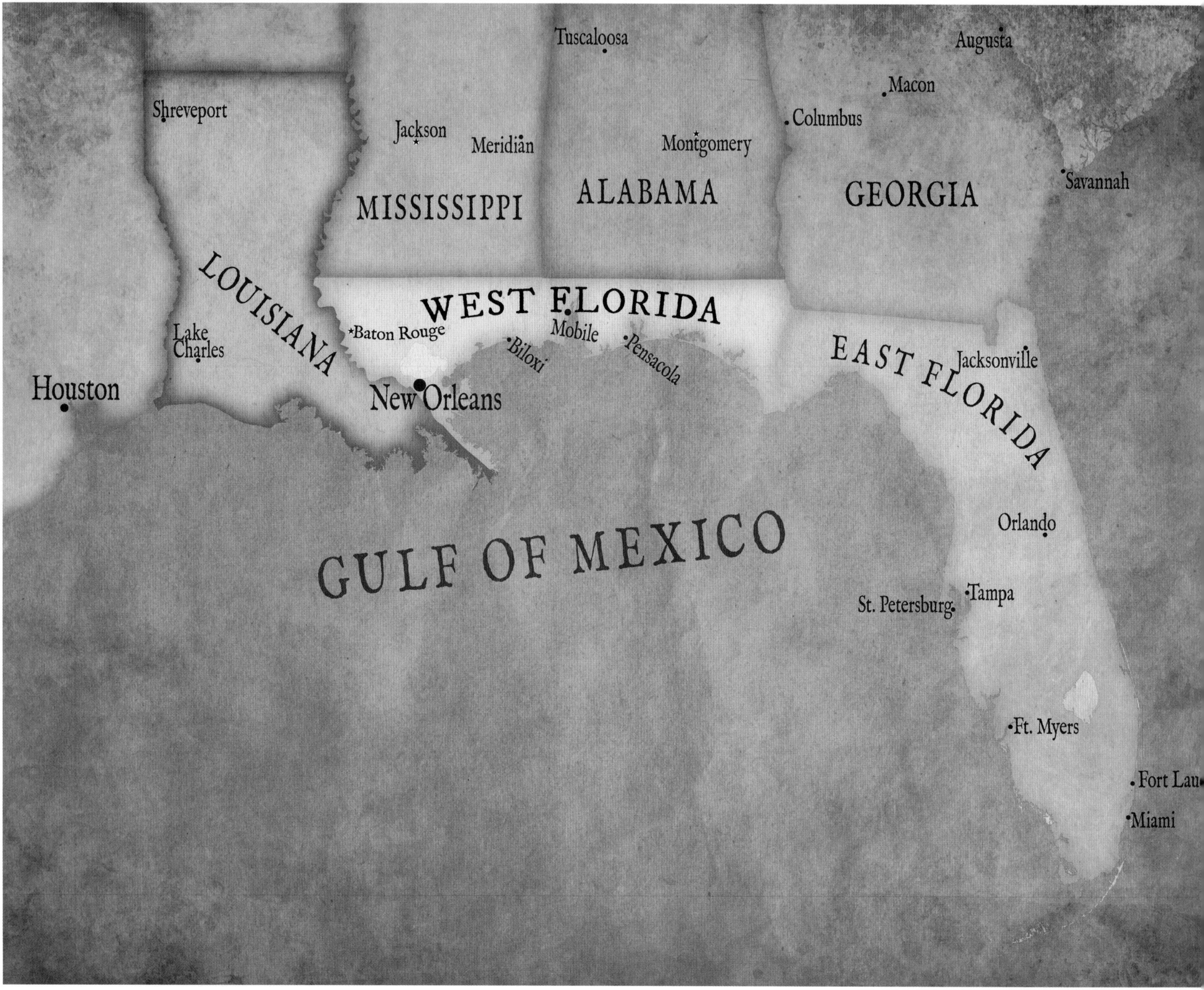

Shreveport

Jackson

Meridian

Tuscaloosa

Augusta

Macon

Columbus

Montgomery

MISSISSIPPI

ALABAMA

GEORGIA

Savannah

LOUISIANA

WEST FLORIDA

Baton Rouge

Mobile

Pensacola

EAST FLORIDA

Jacksonville

Lake Charles

Biloxi

Houston

New Orleans

Orlando

GULF OF MEXICO

St. Petersburg

Tampa

Ft. Myers

Fort Lau

Miami

Including New Orleans. Seems Everyone Wants to Party Here.

Sorting out the West Florida story isn't easy. First, let's define the region. It starts in the Florida panhandle (near Apalachicola) and extends all the way to the Mississippi River. So West Florida includes parts of modern-day Florida, Alabama, Mississippi, and Louisiana, including the cities of New Orleans, Mobile, and Pensacola.

What's confusing is that just about every eighteenth-century power claimed this land. The Spanish, French, British, and Americans all wanted West Florida. I don't know why everyone made such a fuss. Sure, it's the perfect place for spring-break hijinks, but the weekly hurricanes really bring down property values.

Skipping over a lot of messy details, the key date in West Florida history was December 10, 1810, when Americans in the region shot a few Spanish soldiers and flew the flag of the Free and Independent Republic of West Florida. Don't let the "free and independent" tag confuse you; these folks were angling for statehood. But just seventy-four days later, the Spanish reestablished rule. Eventually, the Americans permanently kicked out the Spanish and grabbed the region, but by then West Florida was no longer a unified territory.

The question that remains today is whether the Florida panhandle makes sense as a part of Florida. It's a nine-hundred-mile drive from one end of the state to the other. Granted, I'm including the Florida Keys in that calculation, which may be unfair. But consider this: Pensacola, Florida, is closer to Evansville, Indiana, than it is to Miami. Really. Check it out.

My family used to vacation in Pensacola, and the region seemed more like Alabama or Mississippi than Florida. It took me a long time to figure out why Pensacolans kept talking about tin, until I learned that "tin" is the number that follows nine. No one speaks that way in Orlando or Miami.

OPPOSITE

Republic of West Florida, superimposed on the states as they now exist.

BELOW

Pensacola, Florida, still calls itself the city of five flags, for the five countries that have claimed the city. (I mentioned only four above, but they also count the Confederacy).

★ WEST KANSAS ★

Oil and Water Don't Mix.

Did you ever notice that Kansas, Nebraska, and the Dakotas are wide, horizontal states? That's very much on purpose, because making these states wide (rather than tall) solves a geographic problem.

You see, eastern Kansas gets rain. Western Kansas does not. That means that the western parts of Kansas always had a challenge growing crops, and so fewer people settled there. It was the same story in Nebraska and the Dakotas.

Congress figured that these western netherlands couldn't really make it on their own, and so they connected them to regions with more rainfall. Hence the long, wide states. But anytime you attach regions with different economies, trouble can arise. And so it did in 1992.

Kansas's eastern cities offered tax breaks to companies that would expand in places like Kansas City and Topeka. The technique is effective in attracting new urban industry, but it means somebody else has to pick up the slack. Residents in the west concluded that they would be shouldering the bigger tax burden, without enjoying any of the benefits. That's because western Kansas has lots of oil and natural gas, which is easy to tax. Over time, westerners became increasingly frustrated with the eastern flow of tax dollars. "Witchita and Topeka just want our money," was a common refrain.

So they threatened secession.

The plan was to form a new state, "West Kansas," from the southwest portion of the state. A nonbinding referendum showed that a stunning 84 percent of residents in eight southwest counties were in favor of secession. The breakaway state held a convention in Ulysses in September 1992, adding to the fervor.

Of course, eastern Kansans didn't like the idea at all, and they blocked the plan.

Westerners fought back with a more modern approach to redress of grievances—a dozen lawsuits to upgrade their second-class status.

And thus statehood was put on hold.

OPPOSITE

Map of West Kansas. How many counties were included in the proposal? It depends on your source. This map makes an inference based on published reports. At one point, even counties in neighboring states talked of joining the proposal. The Oklahoma and Texas panhandles, for example, share much more in common with west Kansas than they do with Dallas or Oklahoma City.

BELOW

West Kansas rig pumping out tax revenue (and oil).

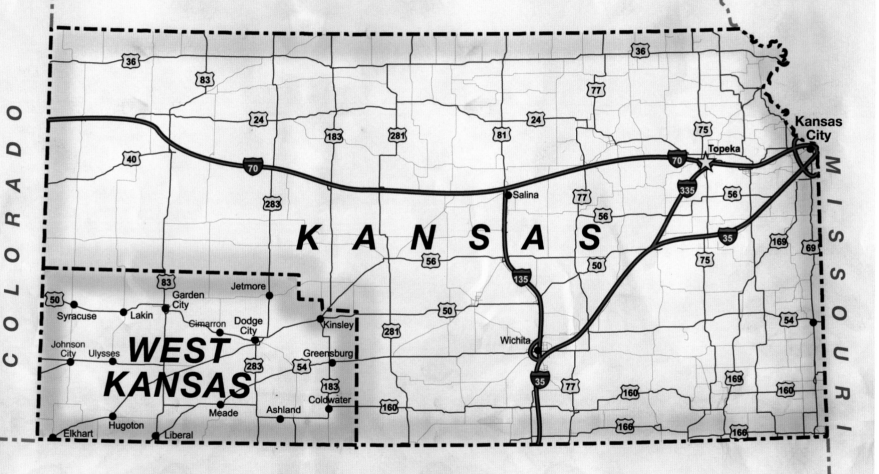

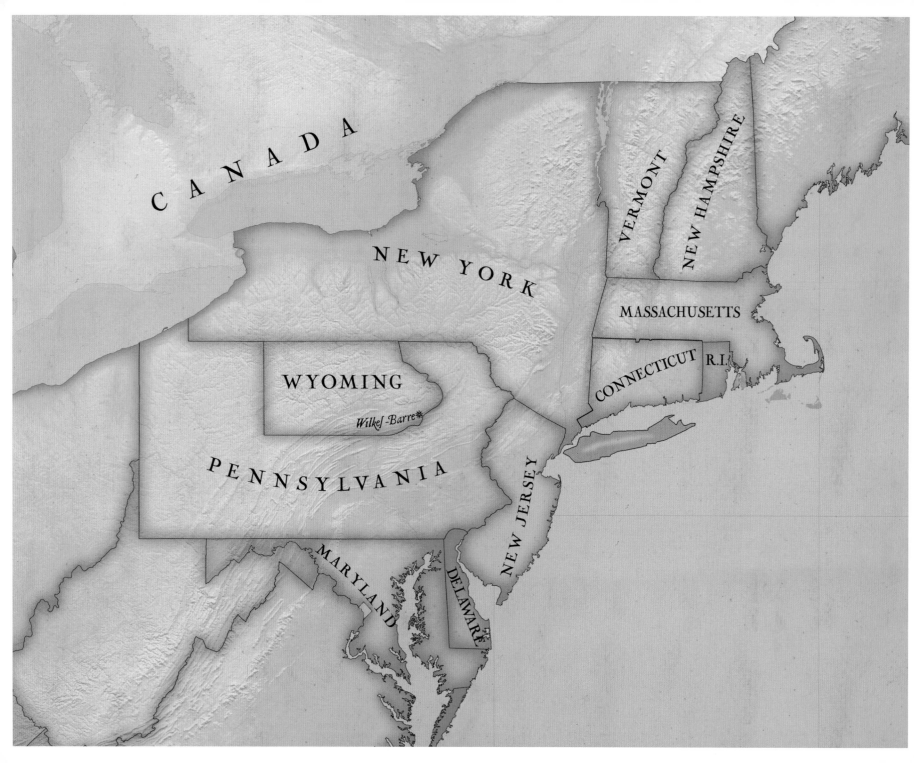

★ WYOMING ★

The Other One. Between New York and Pennsylvania.

What is Wyoming doing in the middle of Pennsylvania? Back in the mid-1600s, Connecticut claimed a huge swath of land stretching all the way to the Pacific Ocean (after hopscotching New York and New Jersey). Connecticut land developers, being an enterprising group, began selling off the choicest part of this real estate, known as the Wyoming Valley. But there was just one problem. The land they were selling was in Pennsylvania, not Connecticut—at least, that's how Pennsylvanians saw it. The Pennsylvania government grew especially upset when Connecticut started establishing cities along the Susquehanna River. Wilkes-Barre, for example, was originally chartered as a Connecticut city. Nowadays, of course, it's in the heart of the Keystone State.

This conflict, called the Yankee-Pennamite Wars, endured for decades—and got pretty testy. At one point, Pennsylvanians brandished weapons and killed some of the Connecticut settlers. After the American Revolution, Congress settled the matter in favor of Pennsylvania.

Had Congress voted the other way, however, the Wyoming Valley would have become part of Connecticut. But given the area's loca-

tion, it might have angled to become a separate state. In fact, there is pretty good evidence that the idea of statehood cropped up as early as 1786. A constitution was drawn up; a governor was named, and Ethan Allen was put in charge of the military. (Yes, that Ethan Allen. And no, he never made furniture.)

After being kicked out of Pennsylvania, Connecticut continued to claim ownership of their coast-to-coast swath, focusing on portions farther west, in what is now Ohio. Eventually, they were kicked out of Ohio, too. Who knows, maybe Connecticut will rekindle its claim and try to grab land even farther west. But Salt Lake City, Connecticut, just doesn't sound right.

OPPOSITE --
Wyoming, superimposed on modern state lines. Instead of being called "Wyoming," it's just as likely that the new state would have been called "Westmoreland," after the Connecticut county that was the previous home of many of the settlers.

RIGHT ---
Connecticut originally claimed a swath of land extending from one ocean to the other. Of course, no one had a clue about just how far that was.

Even after the Pennsylvania matter was settled, Connecticut continued to claim parts of what is now Ohio. Ever heard of Case Western Reserve University? Its name refers to the western "reserve" lands of Connecticut.

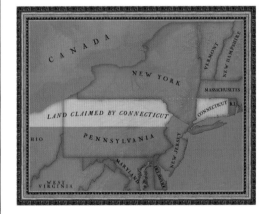

★ YAZOO ★

Government Scandal in Its Formative Years.

We all know that certain politicians are quite willing to sell off political favors. But the Yazoo land scandal took government corruption to a whole new level.

The scam played out as follows: In the late 1700s, Georgia's territory extended all the way to the Mississippi River, including land that now makes up the states of Alabama and Mississippi. Because of the lack of settlers in the western regions, legislators sought to encourage homesteading by drafting a law designed to give away land in two-hundred-acre increments: one plot per household. But four companies were given a sweetheart deal that would make even Halliburton blush. Rather than two hundred acres, they got forty million.

It gets better.

The companies that received these huge tracts of nearly free land were partially owned by—you guessed it—the very politicians who approved the deal.

They thought they'd get away with it, but word leaked about the so-called Yazoo Land Scandal. The people of Georgia were so angry that they voted the offenders out of office.

Imagine that.

Today, of course, graft and profiteering aren't enough to get incumbents voted out. They'd have to commit a more serious crime, such as changing their minds on an issue.

Had the Georgia legislators of the 1780s and '90s been as crafty as modern politicians, the Yazoo land deals might have resulted in an earlier settlement of the region that is now Mississippi. The resulting state lines would likely have followed the borders of the Yazoo tracts. Of course, we can't know exactly how those borders would have eventually been settled, but one thing is certain: "Yazoo" would have been the coolest state-name in the nation. Then again, the word reputedly means "River of Death" in the language of local Native American tribes—probably not the best slogan for travel brochures.

OPPOSITE

The Yazoo land sales had several iterations over the years. This map combines the two biggest tracts (from the 1795 deals) into one state. It might seem incredible that the government would hand over state-size chunks of land to private companies, but such practices were not that unusual. When the time came to build railroads through the West, the U.S. government did exactly the same thing.

BELOW

Yazoo failed one hundred years before the invention of the automobile, so this license plate is a bit of conjecture. But it sure would be interesting to have slogans that reflect the truth a bit more.

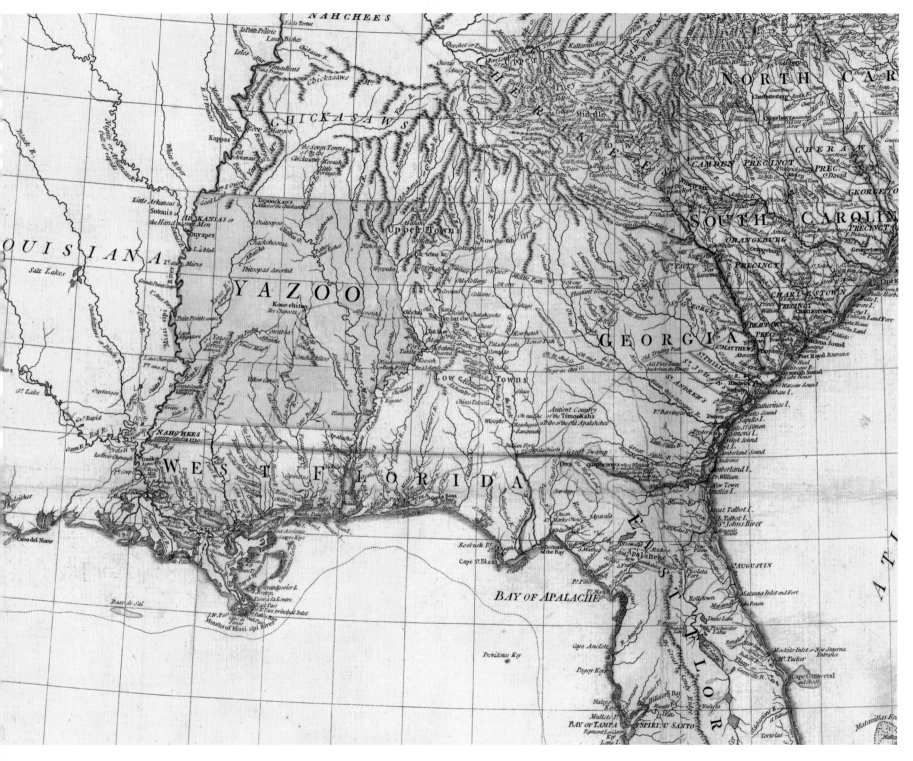

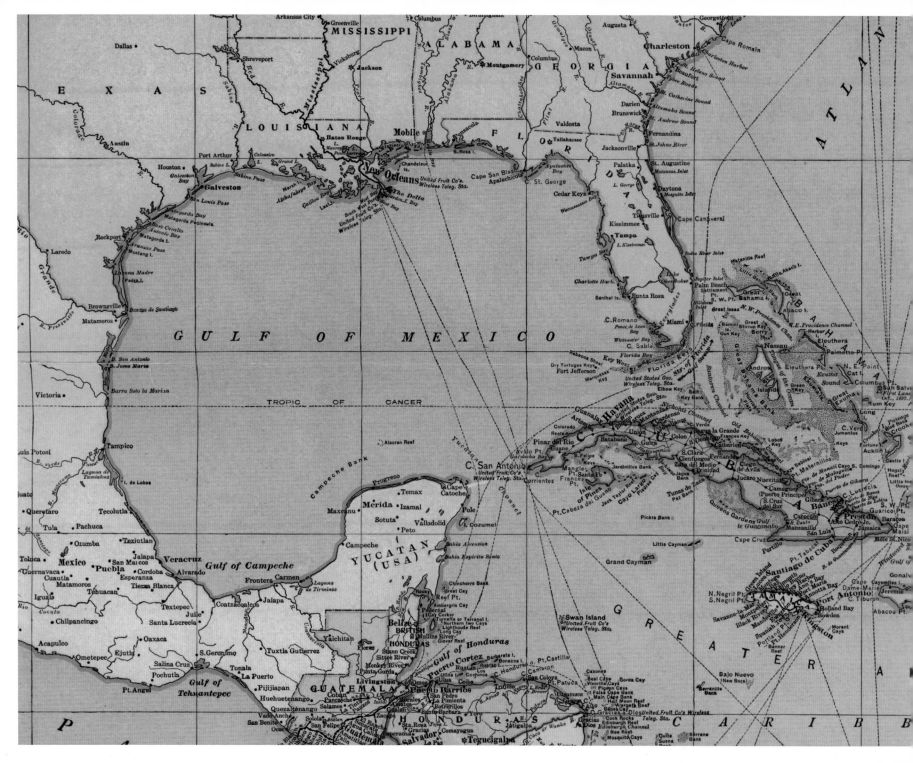

★ YUCATAN ★

Check Out the Hard Rock Café. Oh, and Those Mayan Ruins, Too.

Every year, Mexico's Yucatan peninsula attracts thousands of U.S. honeymooners looking for an authentic American experience. Yes, you read that correctly. How else can you explain why the area's biggest city, Cancun, likes to promote its Hard Rock Café, Olive Garden, and three Wal-Marts? Mexican restaurants are actually hard to find.

So it's no surprise that Yucatan was once proposed as an American state. The idea dates back to the administration of President James Polk in the 1840s, and perhaps even earlier.

The reason was strategic. The Yucatan peninsula juts out into the Gulf of Mexico, much as Florida does. American leaders were worried that the peninsula would fall into the hands of a powerful European nation, which might block American ships from exiting the gulf. So the proposal was that the United States should grab Yucatan before someone else did.

Adding fuel to the fire was Yucatan's ruling class, who wanted to be a part of the United States. In fact, to facilitate annexation, Yucatan declared its independence from Mexico in 1847.

The idea had many powerful American proponents, namely, Mississippi senator Jefferson Davis. President Polk was open to the idea because of Yucatan's "vicinity to Cuba, to the Cape of Florida, to New Orleans, and indeed to our whole southwestern coast; it would be dangerous to our peace and security if it should become a colony of any European power."

But it never came to be. And the reason was racism.

In fact, the United States might have annexed all of Mexico had it not been for a prevailing dislike of nonwhite peoples. Many in Congress at the time unabashedly expressed their desire to keep America as white as possible. I'd quote the example of congressman James Root of Ohio, but his comments are so inflammatory, I can't bring myself to put them in print.

And so Yucatan's statehood drive was dashed. The region reverted to Mexico.

OPPOSITE --

This map shows how Yucatan might have looked as a U.S. territory. Visible are the shipping lanes that were such a big concern.

RIGHT --

Like a zillion other American honeymooners, my wife and I climbed this icon of the Yucatán at Chichén Itzá. The top chamber is empty. Perhaps they should install a Starbucks.

Bibliography

Absaroka "Talk of 49th State Grows in West, Where Dissatisfied Counties Plan to Secede, Add Absaroka to Union," *Hartford Courant*, 3 Mar 1935. "Federal Official Sees Small Prospect of Adding Another Star to U.S. Flag," *Bismark Tribune*, 21 May 1935. Kirk Johnson, "Frontier Spirit Lives On in Breakaway U.S. 'State,'" *New York Times*, 24 Jul 2008.

Acadia Paul Carrier, "Bill Calls for Close Look at Secession," *Portland Press Herald*, 2 Mar 2005. Sarah Schweitzer, "Maine's Split Personality," *Boston Globe*, 27 Feb 2005. Dixon Ryan Fox, *Harper's Atlas of American History* (New York: Harper & Brothers Publishers, 1920).

Adelsverein Moritz Tiling, *History of the German Element in Texas* (Houston: Moritz Tiling, 1913).

Albania Craig S. Smith, "Pro-U.S. Albania Set to Roll Out the Red Carpet for Bush," *International Herald Tribune*, 8 Jun 2007. Craig S. Smith, "For One Visit, Bush Will Feel Pro-U.S. Glow," *New York Times*, 9 Jun 2007. Andrew Purvis, "Albania: 'Please Occupy Us!'" *Time*, 10 Jun 2007.

Alberta and British Columbia Lansing Lamont, *Breakup: The Coming End of Canada and the Stakes for America* (New York: W. W. Norton, 1994). Steven Pearlstein, "O Canada! A National Swan Song?; U.S. Economic, Cultural Weight Threaten Nation's Identity," *Washington Post*, 5 Sep 2000. Steven Pearlstein, "The Eternal Question: Will Canada Survive?" *Toronto Star*, 9 Sep 2000.

Baja Arizona "Debate: Baja Arizona or Bajahaha," *Casa Grande (Ariz.) Dispatch*, 3 Jun 1987. "Baja Arizona Moves to Secede," *The News* (Frederick, Md.), 30 May 1987.

Boston "Brennan Would Make Boston 49th State of the Union," *Boston Daily Globe*, 24 Jul 1919.

Charlotina George Henry Alden, *New Governments West of the Alleghanies before 1780* (Madison: University of Wisconsin, 1897). Clarence Walworth Alvord and Clarence Edwin Carter, *The Critical Period 1763–1765*. (Springfield: Illinois State Historical Library, 1915).

Chesapeake "Secession Legislation Filed for Eastern Shore," *The News* (Frederick, Md.), 1 March 1973. Val Hymes, "The Sandbox Syndrome," *The News* (Frederick, Md.), 15 Mar 1973. "Eastern Shore State Hosts Secession Fete," *Evening Capital* (Annapolis, Md.) 7 Mar 1973.

Chicago "Chicago's Secession Move Stirs Illinois," *New York Times*, 8 Aug 1925. "Chicago Can't Secede," *Appleton (Wis.) Post-Crescent*, 2 Jul 1925. "A New Federation," *Decatur Review*, 21 Aug 1925. Brian Jackson, "Chicago Statehood a Risky Idea," *Chicago Sun Times*, 23 Apr 1995.

Chippewa "Another Hyperborean Territory," *Daily News* (Newport), 12 May 1860.

Comancheria "The Comancheria, Lost Homeland of a Warrior Tribe," *Strange Maps*, http://strangemaps.wordpress.com (accessed 14 May 2007).

Cuba "Statehood for Cuba," *The News* (Marshall, Mich.), 20 Jun 1902. Christopher Ogden, "What About Statehood for—Cuba?" *Time*, 30 Nov 1998. "The Fate of Cuba Discussed by Eminent Americans," *New York Times*, 30 Sep 1906.

Dakota Mark Singer, "True North: Is a State's Name Giving It a Bad Rap?" *The New Yorker*, 18 Feb 2002. Tom Kuntz, "The News Isn't So Hot Here, So Let's Try (North) Dakota," *New York Times*, 26 Aug 2001.

Deseret Territory of Utah. Political History of Nevada, Department of Cultural Affairs, Nevada State Library and Archives, http://nevadaculture.org/docs/nsla/archives/political/historical/hist04.htm (accessed 31 Jul 2008).

England Kermit Holt, "Britons Shy at Idea of Being Our 49th State," *Chicago Daily Tribune*, 4 Mar 1947. "New Talk of Unity," *Time*, 14 Jan 1957.

Forgottonia J. D. Nowlan, "From Lincoln to Forgottonia," *Illinois Issues* 24, no. 9 (1998):27–30.

Franklin Samuel Cole Williams, *History of the Lost State of Franklin* (Johnson City, Tenn.: Overmountain Press, 1993).

Greenland John J. Miller, "Let's Buy Greenland," *National Review*, 7 May 2001. "Deepfreeze Defense," *Time*, 27 Jan 1947.

Guyana "The Mass Emigration from Guyana to the USA," *Guyana, USA* http://www.guyanausa.org

Half-Breed "The Half-Breed Tract in Minnesota," *Chicago Press and Tribune*, 3 Aug 1858. Acts and Resolutions Passed at the First Session of the General Assembly of the State of Iowa. 1846.

Hazard Alden, *New Governments*.

Howland James D. Hague, "Our Equatorial Islands with an Account of Some Personal Experiences," *Century Magazine* 64, no. 5 (Sep 1902). "Hui Panal 'au, Real Life Kamehameha Schools Survivors," *Kamehameha Schools Archives* (Honolulu), http://kapalama.ksbe.edu/archivesU.S. *Insular Areas Application of the U.S. Constitution* (Washington D.C.: U.S. General Accounting Office, 1997).

Iceland "Iceland Seen as 49th State by Congressman," *Los Angeles Times*, 15 Jul 1945. Statehood for Iceland," *Times Recorder* (Zanesville, Ohio), 4 Oct 1945. "U.S. Urged to Offer Statehood to Iceland," *San Antonio Express*, 30 Oct 1945.

Jacinto Weston Joseph McConnell, *Social Cleavages in Texas: A Study of the Proposed Division of the State* (New York: Columbia University, 1925).

Jefferson James T. Rock, *The State of Jefferson: The Dream Lives On!* (Yreka, Calif.: Siskiyou County Museum, 1999). Walt Stafford, "Secession! Rebels Seek to Carve New 'State of Jefferson' Out of Oregon, California," *Gastonia (N.C.) Daily Gazette*, 12 Dec 1941. "Gable's Gold Coast," *Time*, 4 Apr 1938. Bernita Tickner and Gail Fiorini-Jenner, *The State of Jefferson* (Mount Pleasant, S.C.: Arcadia Publishing, 2006).

Lincoln Herb Robinson, "Say Goodbye to Eastern Washington," *Seattle Times*, 22 Feb 1991. Rachel La Corte, "Some GOP Senators Want Eastern Washington as a State unto Itself," *Seattle Times*, 22 Feb 2005. "Some Say East Side Should Be 51st State," *Spokesman Review* (Spokane, Wash.), 23 Feb 2005. Gary Alden Smith, *State and National Boundaries of the United States* (Jefferson, N.C.: McFarland, 2004).

Long Island "State of Long Island," *New York Times*, 11 Jan 1896. Rick Brand, "Long Island: The 51st State," Newsday, 27 Mar 2008. Hank Russell, "Should Long Island Become Its Own State?" *Suffolk Life*, 2 Apr 2008.

Lost Dakota Smith, *State and National Boundaries*.

Lower California Charles O. Paullin, *Atlas of the Historical Geography of the United States* (Washington D.C. and New York: Carnegie Institution of Washington and the American Geographical Society, 1932).

McDonald "Stepchild County Asks Transfer to Kansas," *Kansas City Times*, 23 Feb 1962. "County Receives 'Invasion' Threat," *Great Bend (Kans.) Daily Tribune*, 17 Apr 1961. "Missouri County in 'Secession' Move," *Oshkosh Daily Northwestern*, 11 Apr 1961. Robert Pearam, "Some Towns Find Attractions That Tickle a Tourist's Fancy," *Southern Illinoisan*, 8 May 1964.

Minnesota Smith, *State and National Boundaries*.

Montezuma "New-Mexico as a State," New York Times, 30 Apr 1874. "Democrats Control Senate Situation," *New York Times*, 24 Feb 1903. "Fifty Years of Statehood," H*obbs (New Mexico) Daily News Sun*, 10 Apr 1961. Calvin Horn, "Statehood Struggle Took 60 Years," *Albuquerque Journal*, 2 Feb 1976.

Muskogee Lyle McAlister, "William Augustus Bowles and the State of Muskogee," in *Florida Historical Quarterly* 40, no. 4 (July 1962). Harris Chappell, *Georgia History Stories* (New York: Silver, Burdett, 1905).

Nataqua James Thomas Butler, Isaac Roop, (Janesville, Calif.: High Desert Press, 1994).

Navajo "Statehood Eyed by Navajos," *Albuquerque Tribune*, 27 Apr 1974. "Navajo Chairman Talks of Statehood," *The Independent* (Gallup, N. Mex.), 26 Apr 1974.

Navassa "The American Guano," *Daily News* (Newport). 12 May 1860. Author, "Riot on Navassa Island," *New York Times*, 20 Sep 1889. "Navassa Rioters Sentenced," *New York Times*, 21 Feb 1890.

New Connecticut Hiland Hall, *History of Vermont* (Albany, N.Y.: Joel Munsell, 1868). Walter Hill Crockett, *Vermont: Green Mountain State* (New York: Century History Co., 1921).

New Sweden Amandus Johnson, *Swedish Settlements on the Delaware* (New York: University of Pennsylvania, 1911). Hilde Heun Kagan, ed., *American Heritage Pictorial Atlas of United States History* (New York: American Heritage Publishing, 1966).

New York City "Should New York City Be the 51st State?" *Time*, 21 Jun 1971. Jennifer Senior, "The Independent Republic of New York," *New York Magazine*, 9 Aug 2004.

Newfoundland "How to Be Annexed," *Chicago Daily Tribune*, 30 Sep 1947. "The Way to Statehood," *Chicago Daily Tribune*, 11 Oct 1947. "Here's Your Hat," *Chicago Daily Tribune*, 19 Oct 1947.

Nickajack "A Lincoln Man," *Time*, 21 Feb 1964. Don Dodd and Amy Bartlett-Dodd. *Free State of Winston* (Charleston, S.C.: Arcadia, 2000). Albert James Pickett, *History of Alabama and Incidentally of Georgia and Mississippi* (Charleston, S.C.: Walker and James, 1900).

No Man's Land Morris L. Wardell, "The History of No-Man's Land, or Old Beaver County," *Chronicles of Oklahoma* 1, no. 1 (Jan 1921):60. T. E. Beck, "Cimarron Territory," *Chronicles of Oklahoma* 7, no. 2 (Jun 1929):168

North Slope Wesley Pruden, "Continental Fever for Splithood," *Washington Times*, 24 Jun 1992. Marilee Enge, "North Slope Leaders Talking Secession," *Anchorage Daily News*, 23 Jun 1992.

Panama "Should Panama?" *Parade*, 7 Jun 1964

Philippines "Statehood in Philippines," *Washington Post*, 21 Dec 1900. "Demands Freedom for Philippines: Representative Borland Sees Statehood as Alternative," *New York Times*, 21 Mar 1915.

Popham William Frederick Poole, et al., *The Popham Colony: A Discussion of Its Historical Claims*. (Boston: J. K. Wiggin and Lunt, 1866).

Potomac Carrie Johnson, "The Day Washington and the Suburbs Joined to Become Our 51st State," *Washington Post*, 28 Nov 1979.

Puerto Rico "Jerry Shows 'I'm Still President," *Time*, 17 Jan 1977.

Rio Rico Larry Rohter, "South of the Border Was Once North," *New York Times*, 26 Sep 1987. "Border Town's Story Has More Twists Than Rio Grande," *San Antonio Express*, 20 Jun 2004.

Rough and Ready Hubert Howe Bancroft, *History of California* (San Francisco: History Company, 1886).

Saipan "Combined Guam, CNMI Could Become US State," *Australia Broadcasting Corporation radio report*, http://www.radioaustralia.net.au30 May 2008)Rebecca Clarren, "Paradise Lost," *Ms.* (Spring 2006).

Sequoyah "Statehood Questions," *Manitoba Free Press*. 1 Sep 1905. Serial Set 4912, 59th Congress, 1st sess., Senate doc. 143, p. 87. "The Stillborn State of Sequoyah," Strange Maps, http://strangemaps.wordpress.com (accessed 14 Jul 2007).

Shasta Tom Cameron, "The Tides of Secession," *Los Angeles Times*, 21 Jan 1957. "California: The Second Failure, Business Matters, Etc.," *New York Times*, 2 Jun 1855. "California Hears Cry of Secession," *New York Times*, 9 Dec 1956.

Sicily "The 49th State," *Time*, 15 Apr 1946. Monty Finkelstein, *Separatism, the Allies, and the Mafia* (Bethlehem, Pa.: Lehigh University Press, 1998).

Sonora William Vincent Wells, *Walker's Expedition to Nicaragua* (New York: Stringer and Townsend, 1856). "The Walker Expedition Against Nicaragua," *New York Times*, 15 Oct 1857.

South California Michael Di Leo and Eleanor Smith, *Two Californias: The Truth about the Split-State Movement* (Covelo, Calif.: Island Press, 1983). "LBJ Ignored State's Split," Daily Review (Hayward, Calif.), 23 Jan 1966. Michael McCabe, "27 Counties Backed Idea of a 51st State," *San Francisco Chronicle*, 4 Jun 1992. Jerry Gilliam, "Separatists Push for Bill to Split California into Three States," *Stars and Stripes*, 9 Jun 1993. Maria Goodavage, "California Looks at Splitting Up," USA Today, 14 Apr 1992. Dan Frost, "The New Civil Wars: Secession in the 1990s," *American Demographics*, Mar 1992. Phil Reeves, "North California Dreams of a Split," *The Independent* (London), 5 Jun 1992.

South Florida Ron Littlepage, "Let South Florida Secede and Take Its High Taxes," *Florida Times-Union*, 9 May 2008. "South Florida Wants to Be the 51st State," *Pensito Review*, 9 May 2008.

South Jersey Anne McGrath, "The Near-State of South Jersey," *Associated Press*, 10 Feb 1986. Gregory Byrnes, "Free State of South Jersey Is Really a State of Mind," *Philadelphia Inquirer*, 18 Jan 1982.

South Texas "Texas Threat," *Time*, 26 May 1930. 28th Congress, Sess. II, Res. 9, 10, p. 797+. "Two-State Idea for Texas Grows from Start as Legislature Joke," *Albuquerque Journal*, 9 Apr 1969. "New Tidelands Battle Plan: Divide State," *Galveston Daily News*, 8 Jun 1950.

State X Paullin, *Atlas of Historical Geography. Charles Campbell, History of the Colony and Ancient Dominion of Virginia* (Philadelphia: J. B. Lippincott, 1860).

Superior Jon Lowell, "Superiority Complex," *Newsweek*, 1 Sep 1975. Beth Gauper, "Marquette Rough-Cut Gem of U.P. Michigan" *Pittsburgh Post-Gazette*, 18 Jan 1998. Louise P. Kellogg, "The Disputed Michigan-Wisconsin Boundary," *Wisconsin Magazine of History* 1, (1917–18):304–7.

Sylvania Paullin, *Atlas of Historical Geography. An Ordinance for the Government of the Territory of the United States, North-West of the River Ohio*, 1787.

Texlahoma "Texlahoma Is Given Little Chance of Being 49th State," *Bismark Tribune*, 21 May 1935. "Proposal Made for 49th State," Hammond Times, 21 Jun 1935. "Proposal Made for 49th State," *Chronicle-Telegram* (Elyria, Ohio), 26 Jun 1935.

Trans-Oconee Christopher J. Floyd. "Trans-Oconee Republic," *New Georgia Encyclopedia*, (Athens: University of Georgia Press, 2005)

Transylvania Dixon Ryan Fox, *Harper's Atlas of American History* (New York: Harper & Brothers, 1920). Alden, *New Governments*. Emilius O. Randall and Daniel J. Ryan. *History of Ohio: The Rise and Progress of an American State* (New York: Century History Company, 1915).

Vandalia Thomas Perkins Abernethy, *Western Lands and the American Revolution* (New York: Russell & Russell), 1937. James Donald Anderson, "Vandalia: The First West Virginia?" *West Virginia History* 40, no. 4, (Summer 1979):375–92. Alden, New Governments.

Washington Thomas Suddes, "An Altered State? Call It Addition by Extraction," *The Plain Dealer* (Cleveland), 8 Apr 2007. P. Cherry, *The Western Reserve and Early Ohio* (Akron, Ohio: R. L. Fouse, 1921).

West Florida Robert Higgs, "The Republic of West Florida: Freedom Fight or Land Grab?" *The Freeman: Ideas in Liberty* 31 (June 2005). Isaac Joslin Cox, *The West Florida Controversy, 1798–1818: A Study in American Diplomacy* (Baltimore: Johns Hopkins Press, 1918).

West Kansas Ned Zeman and Lucy Howard, "Kansas Revolt," *Newsweek*, 20 April 1992. Michael Bates, "High Plains Officials Want to Form 51st State," *Associated Press*, 17 Mar 1992. Matthew Schofield, "Southwest Kansas Votes for Secession," *Kansas City Star*, 8 Apr 1992. "Secession Movement Advances," *Kansas City Star*, 12 Sep 1992.

Wyoming Smith, *State and National Boundaries*.

Yazoo Paullin, *Atlas of Historical Geography. C. Peter Magrath, Yazoo, Law, and Politics in the New Republic* (Providence, R.I.: Brown University Press, 1966).

Yucatan Richard Van Alstyne, *The Rising American Empire* (New York: W. W. Norton, 1974). James K Polk, *The Diary of James K. Polk During His Presidency, 1845 to 1849* (Chicago: McClurg, 1910). U.S. Congress, *Congressional Globe*, 30th Cong., 1st sess., 712.

General Works

Franklin Van Zandt, *Boundaries of the United States and the Several States* (Washington D.C.: U.S. Geological Survey, 1976).

Edward M. Douglas, *Boundaries, Areas, Geographic Centers, and Altitudes of the United States and the Several States* (Washington D.C.: U.S. Department of the Interior, 1932).

John T. Faris, *The Romance of the Boundaries* (New York: Harper & Brothers, 1926).

Hilde Heun Kagan, ed., *American Heritage Pictorial Atlas of United States History* (New York: American Heritage Publishing, 1966).

Gary Alden Smith, State and National Boundaries of the United States (Jefferson, N.C.: McFarland, 2004).

Tom Patterson, U.S. National Park Service.

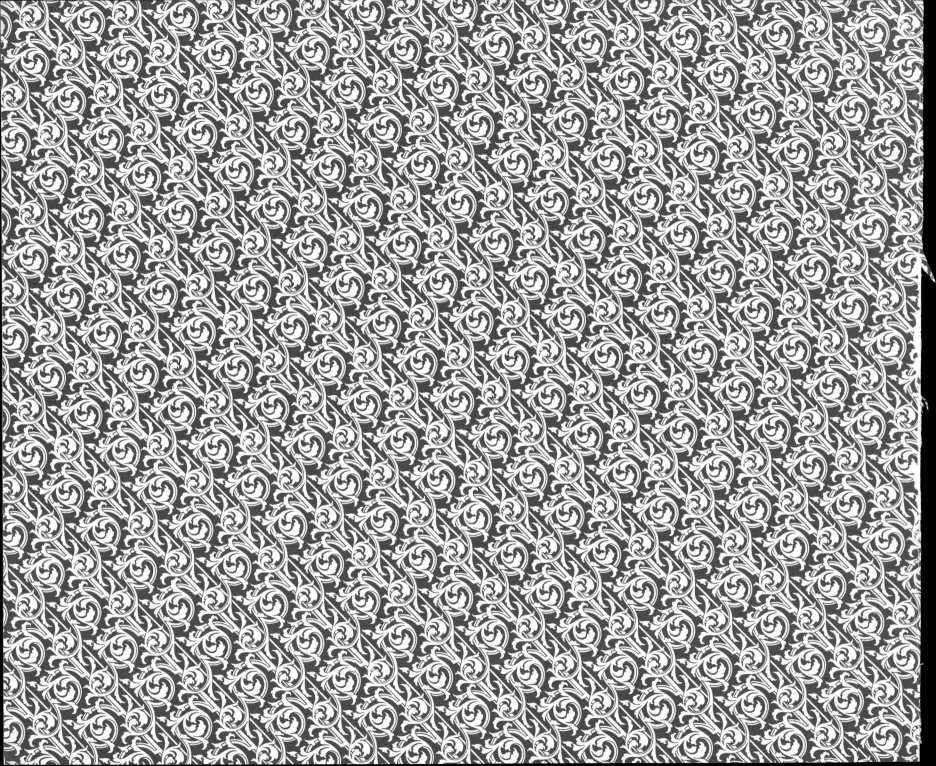